DOT/FAA/AND-740-00/1
DOT-VNTSC-FAA-00-07

Navigation and Landing
Product Team
Washington, DC 20591

Calibration Validation for the New Generation Runway Visual Range System

David C. Burnham
Robert J. Pawlak

Research and Special Programs Administration
John A. Volpe National Transportation Systems Center
Cambridge, MA 02142-1093

Interim Report
July 2000

This document is available to the public
through the National Technical Information
Service, Springfield, Virginia 22161

20000807 068

DTIC QUALITY INSPECTED 4

U.S. Department of Transportation
Federal Aviation Administration

Notice

This document is disseminated under the sponsorship of the Department of Transportation in the interest of information exchange. The United States Government assumes no liability for its contents or use thereof.

Notice

The United States Government does not endorse products or manufacturers. Trade or manufacturers' names appear herein solely because they are considered essential to the objective of this report.

REPORT DOCUMENTATION PAGE		Form Approved OMB No. 0704-0188

Public reporting burden for this collection of information is estimated to average 1 hour per response, including the time for reviewing instructions, searching existing data sources, gathering and maintaining the data needed, and completing and reviewing the collection of information. Send comments regarding this burden estimate or any other aspect of this collection of information, including suggestions for reducing this burden, to Washington Headquarters Services, Directorate for Information Operations and Reports, 1215 Jefferson Davis Highway, Suite 1204, Arlington, VA 22202-4302, and to the Office of Management and Budget, Paperwork Reduction Project (0704-0188), Washington, DC 20503.

1. AGENCY USE ONLY (Leave blank)	2. REPORT DATE July 2000	3. REPORT TYPE AND DATES COVERED Interim Report January 1994-December 1999
4. TITLE AND SUBTITLE Calibration Validation for the New Generation Runway Visual Range System		5. FUNDING NUMBERS A095/A0F090
6. AUTHOR(S) David C. Burnham*, Robert J. Pawlak		
7. PERFORMING ORGANIZATION NAME(S) AND ADDRESS(ES) John A. Volpe National Transportation Systems Center, DTS-53 55 Broadway Cambridge, MA 02142 Scientific and Engineering Solutions, Inc.* 16 Anchor Drive Orleans, MA 02653		8. PERFORMING ORGANIZATION REPORT NUMBER DOT-VNTSC-FAA-00-07
9. SPONSORING/MONITORING AGENCY NAME(S) AND ADDRESS(ES) U.S. Department of Transportation Federal Aviation Administration Navigation and Landing Product Team 800 Independence Avenue, S.W. Washington, DC 20591		10. SPONSORING/MONITORING AGENCY REPORT NUMBER DOT/FAA/AND-740-00/1

11. SUPPLEMENTARY NOTES
*under contract to:
   U.S. Department of Transportation
   Research and Special Programs Administration
   John A. Volpe National Transportation Systems Center
   55 Broadway
   Cambridge, MA 02142		

12a. DISTRIBUTION/AVAILABILITY STATEMENT This document is available to the public through the National Technical Information Service, Springfield, VA 22161		12b. DISTRIBUTION CODE

13. ABSTRACT (Maximum 200 words)

A forward scattermeter, consisting of transmitter and receiver heads mounted on a fork, is used in the New Generation Runway Visual Range (NGRVR) System to assess the clarity of the atmosphere. The scattermeter is calibrated by comparison with reference transmissometers. The consistency of the calibration from one unit to the next is verified by measuring the fork geometry and using a performance model to predict the expected calibration variation for different forks. The model also assesses the possible effects of head variations. The variations in fork geometry through the 650-unit production run are presented. The additional steps needed to complete the calibration validation are defined.

14. SUBJECT TERMS transmissometer, forward scattermeter, runway visual range, RVR, fog			15. NUMBER OF PAGES 98
			16. PRICE CODE
17. SECURITY CLASSIFICATION OF REPORT Unclassified	18. SECURITY CLASSIFICATION OF THIS PAGE Unclassified	19. SECURITY CLASSIFICATION OF ABSTRACT Unclassified	20. LIMITATION OF ABSTRACT

NSN 7540-01-280-5500

Standard Form 298 (Rev. 2-89)
Prescribed by ANSI Std. 239-18
298-102

PREFACE

Ralph Hoar of the Air Force Phillips Laboratory assisted in the Otis tests. Leo Jacobs of the System Resources Corp. participated at all three test sites.

METRIC/ENGLISH CONVERSION FACTORS

ENGLISH TO METRIC

LENGTH (APPROXIMATE)
- 1 inch (in) = 2.5 centimeters (cm)
- 1 foot (ft) = 30 centimeters (cm)
- 1 yard (yd) = 0.9 meter (m)
- 1 mile (mi) = 1.6 kilometers (km)

AREA (APPROXIMATE)
- 1 square inch (sq in, in^2) = 6.5 square centimeters (cm^2)
- 1 square foot (sq ft, ft^2) = 0.09 square meter (m^2)
- 1 square yard (sq yd, yd^2) = 0.8 square meter (m^2)
- 1 square mile (sq mi, mi^2) = 2.6 square kilometers (km^2)
- 1 acre = 0.4 hectare (he) = 4,000 square meters (m^2)

MASS - WEIGHT (APPROXIMATE)
- 1 ounce (oz) = 28 grams (gm)
- 1 pound (lb) = 0.45 kilogram (kg)
- 1 short ton = 2,000 pounds (lb) = 0.9 tonne (t)

VOLUME (APPROXIMATE)
- 1 teaspoon (tsp) = 5 milliliters (ml)
- 1 tablespoon (tbsp) = 15 milliliters (ml)
- 1 fluid ounce (fl oz) = 30 milliliters (ml)
- 1 cup (c) = 0.24 liter (l)
- 1 pint (pt) = 0.47 liter (l)
- 1 quart (qt) = 0.96 liter (l)
- 1 gallon (gal) = 3.8 liters (l)
- 1 cubic foot (cu ft, ft^3) = 0.03 cubic meter (m^3)
- 1 cubic yard (cu yd, yd^3) = 0.76 cubic meter (m^3)

TEMPERATURE (EXACT)
[(x-32)(5/9)] °F = y °C

METRIC TO ENGLISH

LENGTH (APPROXIMATE)
- 1 millimeter (mm) = 0.04 inch (in)
- 1 centimeter (cm) = 0.4 inch (in)
- 1 meter (m) = 3.3 feet (ft)
- 1 meter (m) = 1.1 yards (yd)
- 1 kilometer (km) = 0.6 mile (mi)

AREA (APPROXIMATE)
- 1 square centimeter (cm^2) = 0.16 square inch (sq in, in^2)
- 1 square meter (m^2) = 1.2 square yards (sq yd, yd^2)
- 1 square kilometer (km^2) = 0.4 square mile (sq mi, mi^2)
- 10,000 square meters (m^2) = 1 hectare (ha) = 2.5 acres

MASS - WEIGHT (APPROXIMATE)
- 1 gram (gm) = 0.036 ounce (oz)
- 1 kilogram (kg) = 2.2 pounds (lb)
- 1 tonne (t) = 1,000 kilograms (kg) = 1.1 short tons

VOLUME (APPROXIMATE)
- 1 milliliter (ml) = 0.03 fluid ounce (fl oz)
- 1 liter (l) = 2.1 pints (pt)
- 1 liter (l) = 1.06 quarts (qt)
- 1 liter (l) = 0.26 gallon (gal)
- 1 cubic meter (m^3) = 36 cubic feet (cu ft, ft^3)
- 1 cubic meter (m^3) = 1.3 cubic yards (cu yd, yd^3)

TEMPERATURE (EXACT)
[(9/5) y + 32] °C = x °F

QUICK INCH - CENTIMETER LENGTH CONVERSION

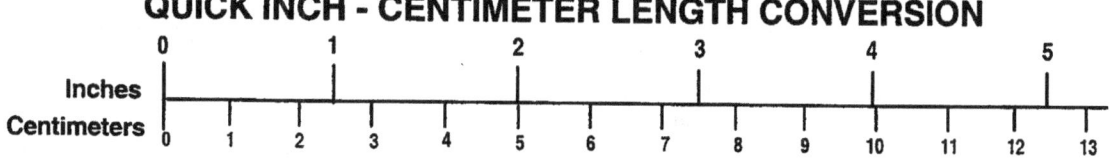

QUICK FAHRENHEIT - CELSIUS TEMPERATURE CONVERSION

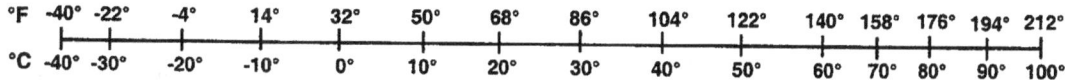

For more exact and or other conversion factors, see NIST Miscellaneous Publication 286, Units of Weights and Measures. Price $2.50 SD Catalog No. C13 10286

Updated 6/17/98

TABLE OF CONTENTS

Section	Page
1. INTRODUCTION	1
1.1 PURPOSE	1
1.2 FORWARD SCATTERMETER ISSUES	1
1.2.1 Characteristics of Visibility Sensor Errors	1
1.2.2 Calibration	2
1.2.3 Calibration Consistency	2
1.2.3.1 Unit-to-Unit	2
1.2.3.2 Different Obstructions to Vision	2
1.2.3.3 Time Variation	3
1.3 RVR SYSTEM ACCURACY REQUIREMENTS	3
1.4 US RVR SYSTEM ACCURACY SPECIFICATION	4
1.4.1 Random Errors	4
1.4.1.1 Current Requirement	4
1.4.1.2 Possible New Requirement	5
1.4.2 Systematic Errors	6
1.4.2.1 Drift	6
1.4.2.2 Unit-to-Unit Variation	6
1.4.2.3 Calibration Device Consistency	7
1.4.2.4 Equal Snow and Fog Response	7
1.4.2.5 Offset	7
1.5 BACKGROUND	8
1.5.1 Scattermeter Development	8
1.5.2 Production History/Validation	8
1.6 SCOPE OF REPORT	9
1.7 BACKGROUND AND ADDITIONAL WORK REQUIRED	10
1.7.1 Unit-to-Unit Calibration Consistency	10
1.7.2 Calibration Simulation Model Validation	11
1.7.2.1 Correct Computer Code	11
1.7.2.2 Completeness	11
1.7.2.3 Correct Scattering Assumptions	11
1.7.2.4 Field Test Validation	11
1.7.3 Calibration Determination	11
1.7.3.1 Calibration Methodology	11
1.7.3.2 Calibration History	12
1.7.4 Calibration Transfer	12
1.7.4.1 Original Procedure	12
1.7.4.2 Interim Procedure	12
1.7.4.3 Long-Term Procedure	12
1.7.5 Calibration Accuracy	12
1.7.6 Long-Term Calibration Maintenance	13

TABLE OF CONTENTS (cont.)

Section	Page
2. FIELD CALIBRATION	15
2.1 TWO RVR SYSTEMS	15
2.2 EARLY PRODUCTION UNITS AT OTIS	15
2.2.1 Relationship of Field Test Results to Calibration Simulation Results	15
2.2.2 Preliminary Fog Calibration	16
2.2.3 Preliminary Calibration Check	16
2.3 GOLDEN UNITS AT OTIS	16
2.4 PRELIMINARY UNITS AT BIRMINGHAM	17
2.5 GOLDEN UNITS AT BIRMINGHAM	17
2.6 MEDIAN FOG CALIBRATIONS	17
2.6.1 First Analysis	17
2.6.2 Second Analysis	18
3. CALIBRATION SIMULATION	21
3.1 FORK GEOMETRY	21
3.2 CMM MEASUREMENT	23
3.3 HEAD GEOMETRY	26
3.4 SIMCAL8 PROGRAM	26
3.4.1 Fixed Parameters	26
3.4.2 CMM Parameters	27
3.5 87 PRODUCTION UNITS	28
3.6 SIMCAL9 PROGRAM	29
3.6.1 Error Correction	29
3.6.2 Data Extraction	30
3.6.3 Results Based on Raw Serial Number	31
3.6.4 Calibration for All Measurements	32
3.6.5 Calibration for 81 Forks	33
3.6.6 CMM Calibration Consistency	33
3.7 PRODUCTION CONSISTENCY	34
3.7.1 Compound Scattering Angle	35
3.7.2 In-Plane Scattering Angle	36
3.7.3 Calibrator Intercepts	36
3.7.4 Receiver Out-Of-Plane Angle	39
3.7.5 Transmitter Out-Of-Plane Angle	39
3.7.6 Head Mount Angles	39

TABLE OF CONTENTS (cont.)

Section	Page
3.7.7 Distances	40
3.7.8 Parameters after Head Corrections	43
3.8 CALCULATED CALIBRATION VERSUS MEASURED PARAMETERS	46
3.8.1 Ratio Volume to Plate Scattering	46
3.8.2 Separate Plate/Volume Scattering	49
3.8.2.1 Compound Angle	50
3.8.2.2 Transmitter-Receiver Distance	51
3.8.2.3 Intercept Differences	51
4. CALIBRATION MODEL IMPROVMENT/VALIDATION	53
4.1 MODEL INCLUDING HEADS	53
4.1.1 Large Receiver Head Alignment Errors	53
4.1.1.1 Unrotated Y-Z Coordinates	53
4.1.1.2 Rotated Y-Z Coordinates	54
4.1.2 Small Head Alignment Errors	57
4.2 PLATE AND FOG SCATTERING PARAMETERS	58
4.2.1 Plate	58
4.2.2 Fog	59
4.2.3 Sensitivity Analysis	59
4.3 WRONG CALIBRATOR LOCATION	60
4.4 FIELD TESTED EARLY PRODUCTION UNITS	61
5. CONCLUSIONS	63
5.1 CALIBRATION MODELING	63
5.1.1 SIMCAL8 Validation	63
5.1.2 Sensitivity to Parameters	63
5.1.3 Comparison to Field Tests	63
5.1.4 Effect of Head Variations	63
5.2 PRODUCTION CONSISTENCY	63
5.2.1 Forks	63
5.2.2 Effect of Heads	63
6. RECOMMENDATIONS	65
APPENDIX A - CMM PROGRAM	A-1
A.1 PROGRAM	A-1
A.1.1 Original	A-1
A.1.2 Added July 1995	A-7
A.2 DISK STORAGE FORMAT	A-8

TABLE OF CONTENTS (cont.)

Section	Page
A.2.1 Original Content	A-8
A.2.2 Added July 1995	A-13
A.3 HARD COPY FORMAT	A-14
A.3.1 Original	A-14
A.3.2 Added July 1995	A-16
APPENDIX B - SIMCAL8 DETAILS	B-1
B.1 HEAD GEOMETRY	B-1
B.2 SCATTERING CALCULATION	B-1
REFERENCES	R-1

LIST OF FIGURES

Figure	Page
Figure 1. Box Plot with Accuracy Analysis	5
Figure 2. Possible New Accuracy Requirement	6
Figure 3. Laser Device for Checking Scattering Angle	9
Figure 4. Otis Fog Calibration Results (1994-97)	17
Figure 5. Fog Calibration of Otis "Deployed" Scattermeters	19
Figure 6. Intercomparison of Early Production Deployed Scattermeters	20
Figure 7. AGI Calibration of Birmingham Scattermeters	20
Figure 8. Scattermeter Drawing: Top and Side Views That Define SIMCAL7 Coordinate System	21
Figure 9. Scattermeter Photographs (side, bottom)	21
Figure 10. Calibration Plate	22
Figure 11. Offset Views of Scattermeter with Calibrator Installed	22
Figure 12. Installing Fork	23
Figure 13. Fork in Place with Calibrator Plate Fixture Installed	23
Figure 14. Receiver Mount Measurements	23
Figure 15. Receiver Head, Close View	23
Figure 16. Manual Setting for Transmitter Mount	24
Figure 17. Calibration Plate Fixture Measurement	25
Figure 18. Transmitter Head Detail	25
Figure 19. Distribution of Calculated Calibrations	28
Figure 20. Relationship Between Assembly and Raw Serial Numbers	31
Figure 21. Results of SIMCAL9 Calculation vs. Raw Serial Number	32
Figure 22. Measurement Date vs Raw Serial Number	32
Figure 23. Figure 22 for 81 Teledyne Forks	33
Figure 24. Distribution of Normalized Ratio of Volume to Plate Scattering	33
Figure 25. Calibrations for Duplicate Measurements	34
Figure 26. Compound Scattering Angle vs Raw Serial Number	35
Figure 27. In-Plane Scattering Angle vs Raw Serial Number	36
Figure 28. Intercept Spacing vs Raw Serial Number	36
Figure 29. Comparison of Two Intercept Difference Values	37
Figure 30. Receiver & Transmitter Calibrator Y Intercept vs Raw Serial Number	37
Figure 31. Receiver & Transmitter Calibrator Y Intercept vs Raw Serial Number	38
Figure 32. Y and Z Calibrator Y Intercept Rx-Tx Differences vs Raw Serial Number	38
Figure 33. Receiver Out-Of-Plane Angle vs Raw Serial Number	39

Figure 34. Transmitter Out-Of-Plane Angle vs Raw Serial Number 39
Figure 35. Fork Mount Angles vs Raw S/N: Top = Θ_r; Middle = Θ_t; Bottom = Φ_r 40
Figure 36. Fork Distances vs Raw S/N: Top = Receiver to Transmitter, Middle = Transmitter to Transmitter Calibrator Intercept; Bottom = Receiver to Receiver Calibrator Intercept 41
Figure 37. Receiver & Transmitter X vs Raw Serial Number 42
Figure 38. Receiver & Transmitter Y vs Raw Serial Number 42
Figure 39. Receiver & Transmitter Z vs Raw Serial Number 43
Figure 40. Comparison of Intercept Differences Before/After Head Corrections 43
Figure 41. Changes In Rx-Tx Intercept Differences from Head Corrections 44
Figure 42. Beam Distances vs Raw S/N: Top = Receiver to Transmitter; Middle = Transmitter to Transmitter Calibrator Intercept; Bottom = Receiver to Receiver Calibrator Intercept 45
Figure 43. Compound & In-Plane Scattering Angle vs NRVP 46
Figure 44. Rx-Tx Calibrator Intercept Difference vs NRVP 46
Figure 45. Receiver & Transmitter Out-Of-Plane Angles vs NRVP 46
Figure 46. Fork Mount Angles vs NRVP: Top = Θ_r; Middle = Θ_t; Bottom = Φ_r 47
Figure 47. Distances vs NRVP: Top = Rx-Tx; Middle = Tx-Cal Int.; Bottom = Rx-Cal Int. 47
Figure 48. Receiver X, Y, Z vs NRVP 48
Figure 49. Transmitter X, Y, Z vs NRVP 48
Figure 50. Intercept Differences vs NRVP 49
Figure 51. Correlation Between Plate and Volume Scattering 49
Figure 52. Correlation Between Number of Plate and Volume Scattering Points 50
Figure 53. Correlation Between Volume/Plate Scattering and Compound Scattering Angle 50
Figure 54. Correlation Between Volume/Plate Scattering and Transmitter-Receiver Distance 51
Figure 55. Dependence of Volume Scattering on Intercept Differences 51
Figure 56. Dependence of Plate Scattering on Intercept Differences 52
Figure 57. Dependence of Plate Scattering on Intercept Differences Rotated by 27 Degrees 52
Figure 58. Dependence of Normalized Number of Points on Y Intercept Changes 53
Figure 59. Dependence of Normalized Number of Points on Z Intercept Changes 53
Figure 60. Effect of Y Intercept Changes 54
Figure 61. Effect of Z Intercept Changes 54
Figure 62. Variation of Mean Scattering Angle with Receiver Intercept Displacement 55
Figure 63. Mean Scattering Angle vs Receiver Intercept Displacement, after Rotation 55
Figure 64. Points vs Y' Displacement 55
Figure 65. Points vs Z' Displacement 55
Figure 66. Effect of Y' Intercept Changes 56
Figure 67. Effect of Z' Intercept Changes 56
Figure 68. Effect of Intercept Displacements on NRVP (top) and Change in NRVP (bottom) 58

Figure 69. Calibration Distribution for Case 3 60
Figure 70. Normalized Scattering Ratio for Displaced Calibrator 61
Figure 71. Pattern for Scanning Ellipse B-1

LIST OF TABLES

Table	Page
Table 1. Visibility Sensor Calibration Errors	2
Table 2. US Visibility Sensor Accuracy Requirements	4
Table 3. Preliminary Calibration Sensors	15
Table 4. Preliminary Calibration Check	16
Table 5. Golden Scattermeters at Otis	16
Table 6. Scattermeters at Birmingham (1994-95)	17
Table 7. Scattermeters at Birmingham (1995-97) and Otis (1998)	17
Table 8. Beam Size (mrad)	27
Table 9. Sample Input Data for SIMCAL8	28
Table 10. Sample Output Data for SIMCAL8	28
Table 11. Sample Output Data for SIMCAL9, Same Cases as Table 10	30
Table 12. Statistics of SIMCAL9 Calculation for 87 Teledyne Cases	30
Table 13. Summary of Parameter Correlation with NRVP	49
Table 14. Results of Sensitivity Analysis	59

1. INTRODUCTION

The Federal Aviation Administration's (FAA) new generation Runway Visual Range (RVR) system uses a forward scattermeter rather than the conventional transmissometer to measure the atmospheric extinction coefficient. Although the first systems were fielded in 1990 to demonstrate system reliability, the system was not approved for national deployment until August 1994 because of the extensive testing and modifications needed to make sure that the new sensor technology meets the operational requirements. The history of this development through August 1996 was documented.[1]

Forward scattermeter calibration was identified long ago[2] as one of the critical issues for the acceptance of forward scattermeters for RVR use and required a major effort in the development process. In contrast to the transmissometer, which can be calibrated in the field, a forward scattermeter can be calibrated only by comparison to a transmissometer. The calibration of units, which have been compared to a transmissometer, then is transferred to other units by means of scattering plates that simulate the scattering from fog. The validity of this simulation depends upon the unit-to-unit consistency of the scattering geometry.

1.1 PURPOSE

The purpose of this report is to document the remaining activities since August 1996 that were needed to:

1. Determine the best fog calibration for the national deployment forward scattermeter, and
2. Verify that the production tolerances were good enough to meet the specified limit (± 7 percent) on unit-to-unit variation in the fog response.

1.2 FORWARD SCATTERMETER ISSUES

The transmissometer was developed in the 1940s and was deployed at airports for use in estimating runway visual range after appropriate testing, but without a formal specification, because it clearly provided more consistent results than human observations. The forward scattermeter was developed in the late 1960s as a research tool for measuring visibility without the strict installation and maintenance requirements of the transmissometer. However, the convenience of the forward scattermeter came at the expense of a more complex and uncertain calibration procedure compared to the transmissometer's.

1.2.1 Characteristics of Visibility Sensor Errors

Visibility sensors (VS) measure the atmospheric extinction coefficient σ. Most calibration errors can be expressed in terms of the slope b and offset a in the equation: $\sigma_{TRUE} = a + b\, \sigma_{MEASURED}$. Table 1 lists the nature of the slope ($b \neq 1.00$) and offset errors ($a \neq 0.00$) for the transmissometer and the forward scattermeter.

Table 1. Visibility Sensor Calibration Errors

Sensor Type	Slope Error	Offset Error
Transmissometer	Little error. Sensor is self calibrating.	Major source of error: 100% light setting
Forward Scattermeter	So many sources of error that calibration is done with scattering plate. Since fog and plate scattering are different, correct fog calibration depends upon consistency of scattering geometry.	Caused by scattered light. Electronic offsets normally small.

1.2.2 Calibration

Each fielded transmissometer can be calibrated absolutely on a clear day. The calibration of a fielded forward scattermeter is indirect:

1. Several forward scattermeters are calibrated by comparing their readings with one or more transmissometers in the same fog.

2. Scattering devices which simulate the scattering in fog then are measured in the calibrated scattermeters.

3. A scattering device then is placed in a fielded scattermeter and the gain adjusted to give the same reading as in the scattermeters that were compared to the transmissometer(s).

Each step in this process can introduce errors. Moreover, the final calibration of the fielded sensor depends upon the assumption that the scattering from the calibration device is equivalent to that from fog.

1.2.3 Calibration Consistency

The forward scattermeter has a greater dynamic range than the transmissometer because the scattered signal is proportional to the atmospheric extinction coefficient (rather than the exponential of the extinction coefficient, which is produced by the transmissometer). However, the consistency of the scattermeter's response can vary (a) from one unit to the next, (b) for different obstructions to vision, and (c) as a function of time.

1.2.3.1 Unit-to-Unit

Unit-to-unit variations in fog calibration were first observed in field tests and were traced to variations in the sensor's scattering geometry. The calibration device (a) occupies a different portion of the scattering volume (typically a plane) than the obstruction to vision and (b) likely has different scattering properties than the obstruction to vision. Consequently, even relatively small variations in the scattering geometry from one unit to the next can prevent the scattering from the scattering device from being proportional to the scattering from the obstruction to vision.

1.2.3.2 Different Obstructions to Vision

Different obstructions to vision can have dramatically different angular distributions of scattering. Since a scattermeter typically measures scattering over a relatively small range of scattering angles, the relationship between the scattered signal and the extinction coefficient can be significantly different for different obstructions to vision.

1.2.3.3 Time Variation

Window contamination and electronic drift can lead to short term (month-by-month) variations in the response of a forward scattermeter. Such sources of drift can be eliminated by periodic window cleaning and recalibration using the scattering device. Over many years (nominal system lifetime is 20 years) other sources of drift are possible. For example, the characteristics of the calibration device might change over time. Subtle changes in scattering geometry might develop (perhaps via rough handling) and be very difficult to detect. Such changes can be detected only by repeating the process originally used to calibrate the forward scattermeter (by comparison to a transmissometer) and to verify its scattering geometry.

1.3 RVR SYSTEM ACCURACY REQUIREMENTS

In principle, the required accuracy for an RVR system should be based on an operational requirements analysis. However, no such analysis has ever been carried out. The International Civil Aviation Organization (ICAO) has defined two RVR accuracy levels — desired and achievable:

1. The desired accuracy is based on the reported RVR increments and specifies an accuracy that is a small fraction of the reporting increment. This requirement means that the reported value is likely to be the actual value. An accuracy based on the reporting increments is logical, but is rather arbitrary because the selection of reported values is itself somewhat arbitrary.

2. The attainable accuracy is based on the performance of instruments for measuring the clarity of the atmosphere. Although RVR usually (i.e., when runway lights are more visible than black objects) depends also on the measurement of the ambient light level and the runway light intensity, these sources of error were not considered in defining attainable accuracy.

An operational requirements analysis would address a number of considerations:

1. RVR is used operationally by defining a minimum RVR value which will safely permit each airport operation that is affected by reduced visibility. *The required accuracy of the RVR assessment should be related to the sensitivity of the safety of the operation to errors in RVR value.*

2. RVR is an estimate of how far the pilot can see down the runway, which depends upon the visibility of the runway lights and markings. Under conditions of reduced RVR, the runway light intensity is normally set high enough that the lights are more visible than the markings. RVR also is used to approximate the visibility that the pilot will experience in the approach to landing on the runway. *An appropriate accuracy goal for an RVR system is that the errors that can be controlled by sensor design be insignificant in comparison to the error sources that cannot be controlled.* An RVR system uses instrumental measurements at one or more locations to predict the subjective experience of a pilot inside an aircraft at another location. This prediction is subject to many uncontrolled factors, including variations in pilot's eyesight and windscreen characteristics; spatial and directional variations in the atmospheric clarity, ambient light level, and runway light intensity; and uncertainties in the equation used to relate the sensitivity of the eye as a function of the ambient light level. A recent analysis[3] used spatial variations in fog density to assess the required accuracy for assessing the atmospheric clarity.

Requiring an RVR system accuracy greater than is necessary to meet realistic operational requirements might result in significantly higher costs and might even result in a system specification that cannot be practically realized.

1.4 US RVR SYSTEM ACCURACY SPECIFICATION

The US RVR accuracy requirement[4] was formulated in 1985 and was based on four considerations:

1. The accuracy should be high enough to provide user confidence in the reported RVR values.

2. The accuracy should be achievable using forward scattermeter technology, which could eliminate the frequent maintenance requirements experienced with the US transmissometer.

3. The tolerance for random errors is greater than for systematic errors. Random sensor errors will be readily masked by natural fluctuations in the atmosphere. The instrument readings should not, however, be biased with respect to the actual characteristics of the atmosphere.

4. Instead of defining a system accuracy specification, separate accuracy requirements were defined for each of the three sensors (extinction coefficient, ambient light level, and runway light intensity). This approach separates the testing requirements for the three sensors and permits practical test methodologies to be defined for assuring sensor compliance with a specification.

1.4.1 Random Errors

1.4.1.1 Current Requirement

The current US visibility sensor accuracy requirements are listed in Table 2 for two extinction coefficient ranges. The RMSE (root-mean-square-equivalent) limit says that, for a Gaussian error distribution, the standard deviation of the measurement must be less than the specified value at the 90%

Table 2. US Visibility Sensor Accuracy Requirements

Extinction Coefficient	MOR	Accuracy
1.5-10 km^{-1}	300-2000 m	20% RMSE
>10 km^{-1}	< 300 m	15% RMSE

confidence limit. Specifically, this requirement means that 90% of the errors must be less than 1.65 times the RMSE accuracy value. The US accuracy criteria are shown as vertical boundaries in box plots (see Figure 1) and are designed to be applied to experimental data using the two-transmissometer US reference standard with a 10% homogeneity criterion. In this accuracy definition, any errors in the reference standard are incorporated into the accuracy requirements. In addition to the RMSE requirement, less than 1% of the measurements in each range may have errors greater than a factor of two.

Figure 1 shows a sample box plot* where the accuracy analysis of Table 1 has been applied. The dense fog "FOG" MOR bin (sum of MOR bins labeled with "F") of Figure 1 represents the second range of Table 1 ($\sigma > 10$ /km) and has been a common feature of Volpe Center box plots for many years. The light fog "LFOG" (sum of MOR bins labeled with "L") has been added to Figure 1 and represents the first range of Table 1 ($1.5 < \sigma < 10$ /km). The percentages along the left of the box plot show what fraction of the points in each MOR bin meet the US accuracy requirement for that bin. Note the asymmetry in the positions of the upper and lower MOR ratio limits in Figure 1; this asymmetry stems from the

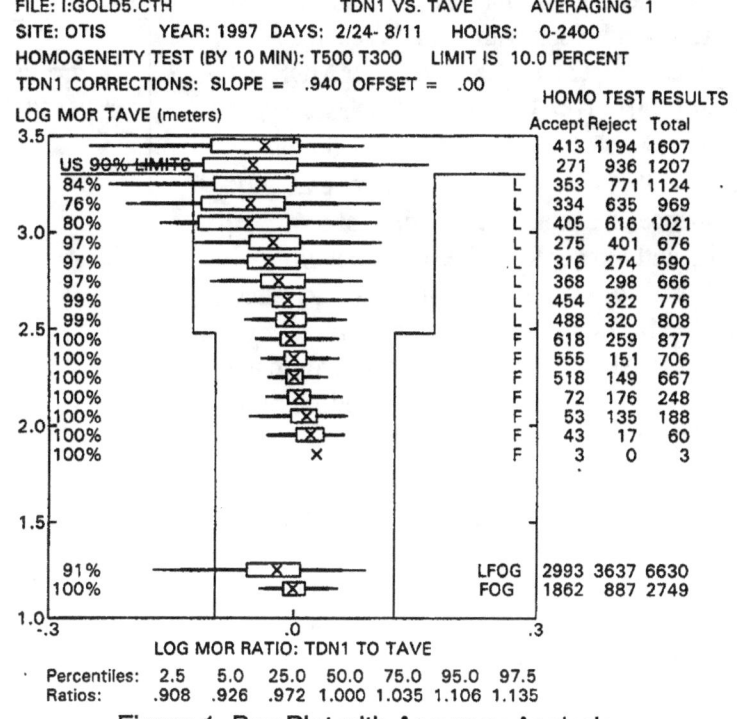

Figure 1. Box Plot with Accuracy Analysis

different fractional effects of large positive and negative percentage errors. Earlier box plots plotted the greater fractional error (i.e., upper MOR ratio limit) as the limit for both upper and lower MOR ratio errors.

The two analysis ranges are summarized at the bottom of Figure 1 in the bins with the "LFOG" and "FOG" labels. The "FOG" bin easily meets the 90% accuracy requirement, but the "LFOG" bin (91 %) just meets the requirement because of a shift in the MOR ratio distribution for bins with LOG MOR above 3.0. Such shifts are typically not observed; the shift in Figure 1 may reflect either a reference transmissometer or a scattermeter anomaly. In any case, the US accuracy requirement is *always* more difficult to meet for the "LFOG" range than for the "FOG" range and the portion of the "LFOG" range above LOG MOR of 3.0 is the most critical.

1.4.1.2 Possible New Requirement

The current requirement is most difficult to meet for MOR above 3000 ft (1000 m) where the US reporting increment is 500 ft (e.g., RVR = 3000 ft, 3500 ft, 4000 ft, etc.), because the RVR is far above the highest RVR minimum of 2400 ft for Category I landings. Because it does not make sense to have the critical performance of the extinction coefficient sensor be for a region where the operational accuracy requirement is relatively unimportant, an alternative accuracy requirement may

---

* The box plot defines logarithmic bins of meteorological range (MOR) based on a reference sensor (in this case, TAVE, the average of the two Otis transmissometers). Then, for each MOR bin, it plots the percentiles of the distribution of MOR ratio of the test sensor (in this case, TDN1) to the reference sensor.

be desirable. For example, Figure 2 shows how Figure 1 would look with a single accuracy range requirement of 15% RMSE for MOR below 1000 m ($\sigma > 3$ /km). For this analysis range, all MOR bins except the top individually meet the 90% confidence level. An overall 90% confidence level could be satisfied for a significantly lower RMSE error, although perhaps not as low as 10% (which would reduce the LOG MOR ratio error band in Figure 2 by one third). Note that in Figure 2, the error in the median (50 percentile) FOG MOR ratio is only 1%. If the sensor under test has the maximum allowed error of 7%, then meeting the 15 percent RMSE requirement would be marginal.

The new accuracy requirement illustrated in Figure 2 might be considered in the 2000 revision of the RVR system specification.

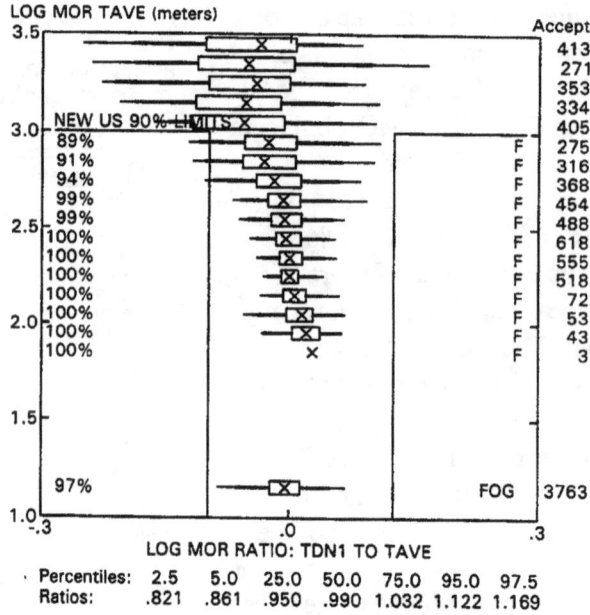

Figure 2. Possible New Accuracy Requirement

1.4.2 Systematic Errors

The basic goal of the original RVR specification was to keep normal systematic errors within the limits of ± 10%.

1.4.2.1 Drift

The drift specification limits calibration variations to ±10% within the 90-day preventive maintenance cycle of the FAA. If a sensor just meets this specification and the contamination build up is constant, then the average error over 90 days would be 5%. Since electronic drift should be small, this specification pertains mostly to the effects of window contamination. Either the windows must be protected enough to limit contamination or the measurement must be corrected for the attenuation caused by window contamination.

Time variations in sensor calibration are readily detected by remeasuring the calibration device.

1.4.2.2 Unit-to-Unit Variation

Unit-to-unit variation in a scattermeter's fog calibration are related to the scattering geometry of each individual sensor and, therefore, are approximately constant over time. The limit on this variation, therefore, was set as a value below (± 7%) the desired 10% systematic error limit.

Testing for unit-to-unit calibration variations is not easily done. Strict application would require fog comparisons for all sensor units to at least a few "reference" scattermeters. Such testing is impractical. The alternative eventually selected about two thirds of the way through the initial production run of VS sensors was to make accurate measurements of the geometry of each sensor fork and then calculate the resulting variations in expected fog calibration. This procedure has not yet been completely validated and will form a major part of this report.

1.4.2.3 Calibration Device Consistency

The scattermeter calibration should not depend significantly on which calibration device is used to set its calibration. The limits on this source of error was set at ± 3%.

Calibrator consistency can be readily checked conducting a round robin measurement of many calibrators in many scattermeters. The measurements must be consistent to within ± 3% of giving the same calibration for each sensor and the same equivalent fog density for each calibration device.

The following two requirements were not in the original 1985 specification, but will be formally added to the 1999 specification to deal with issues that arose during the extensive testing of the US RVR forward scattermeter.

1.4.2.4 Equal Snow and Fog Response

The original design of the US RVR forward scattermeter used a beam-center scattering angle of 35 degrees. Field tests showed that the sensor response to snow was about 30% lower than its response to fog with the same extinction coefficient. Increasing the beam-center scattering angle to 42 degrees reduced the relative fog response more than the snow response to give approximately equal responses to fog and snow. A realistic specification, consistent with the goal of 10% systematic error, would define the "equality" of snow and fog responses as ± 7%.

1.4.2.5 Offset

Scattermeter testing through 1985 uncovered no sensor offsets that would affect RVR measurements. One of the virtues of the forward scattermeter was that it could readily measure high visibilities (e.g., 10 miles; $\sigma = 0.20$ /km) that would require a very long transmissometer baseline. Since the highest RVR value is closer to one mile, even a sensor offset equivalent to 10-mile visibility would have at most a 10-percent effect on accuracy.

Scattermeter offsets can have two sources: self scattering of the transmitted signal and electronic pickup between the transmitter and receiver. The second design for the US RVR forward scattermeter was observed to have large (> 0.5 /km) self-scatter offsets because the beams hit the hoods which protected the windows and the sensor support arms. The self-scatter offsets were eliminated by increasing the diameter of the hoods and painting black all sensor parts that could contribute to self scattering. Unfortunately, field monitoring of the final US RVR scattermeter design showed that a small fraction of the scattermeters exhibited electronic offsets large enough to affect RVR accuracy. The largest observed offset was about 1.0 /km. The offset was traced to a ground loop in the design. The largest simulated offsets in the laboratory were also approximately 1.0 /km. A simple method of eliminating this offset is currently under test. The goal of the RVR offset specification will be to keep the offset within ± 0.2 /km.

Testing for offsets has some of the characteristics of testing for unit-to-unit calibration variations. It is hard to be sure that the specification is met without testing all units. In the case of ground-loop offsets, the problem is compounded by the subtle dependence of the ground loop error on the grounding characteristic of the installation, which are observed to be time dependent. In this case, the most convincing proof of a design free of significant ground loop errors is to model the circuit to understand the offset mechanism and artificially insert the largest possible ground loop offset.

1.5 BACKGROUND

The design and production validation methods varied over the period when the US RVR forward scattermeter was developed and manufactured. These variations pose significant challenges to a satisfactory verification that all scattermeters meet the specifications described in the last section.

1.5.1 Scattermeter Development

The national deployment scattermeter was the third design for the US RVR scattermeter. However, the look-down fork design used was finalized in the second design (early 1993) and was not changed in the final design apart from being painted black instead of white. The scattermeter receiver and transmitter heads of the final design were internally very similar to the original design. They had only minor internal electronic changes; the most significant changes were in the hood and heater designs.

The redesign of the scattering geometry between the first and second designs (change from look-out to look-down geometry) was based on calculations of the relationship between calibration errors and variations in scattering geometry. Calibration simulations[5] enabled the discovery of several principles that lead to more consistent calibrations:

1. Tapered beams are less sensitive to alignment errors than sharp-edged beams. Fortunately, the Handar heads generate tapered beams.

2. The effects of alignment errors are minimized if one beam has a larger footprint on the calibrator plate than the other beam.

3. When one beam footprint is larger than the other, the resulting calibration depends only on the beam diameter consistency of the beam with the larger footprint. Becasue the Handar receiver has a more consistent diameter than the transmitter, the position of the calibrator was moved away from the receiver to make its footprint larger and the transmitter's footprint smaller.

When the sensitivity of the scattering geometry has been reduced by these methods, the primary source of calibration error becomes the exact scattering angle; the scattering from fog varies rapidly with scattering angle. To minimize this source of error, tight production tolerances were placed on scattering angle. The final error analysis (Section 3.8.1, Figure 43) showed that the scattering angle was indeed the most critical parameter of the scattering geometry.

1.5.2 Production History/Validation

The look-down forks are bent from aluminum pipe and then welded to the head flanges using a jig. The first units did not consistently meet the 42.00 ±0.25 degree scattering angle specification and had to be bent into tolerance. The jig then was modified to yield forks meeting the scattering angle specification without adjustment.

The first forks were tested using a laser beam device (see Figure 3) to measure the nominal scattering angle. Subsequently, a coordinate measuring machine (CMM) was used to measure the forks and calculate a number or production control parameters (see Section 3.1). The most significant parameters were the 42.00 ±0.25 degree scattering angle and a maximum displacement of 0.30 inches between the intercepts of the receiver and transmitter beams at the calibrator plate. Hard copy reports were retained from these tests to document compliance with the specification.

The CMM data saved from the original measurement program was not adequate to determine the exact sensor geometry for use in a simulation program. In July 1995, the CMM software was changed to retain all necessary measurements in machine-readable format. The exact sensor geometry can, therefore, be calculated for forks measured after that time.

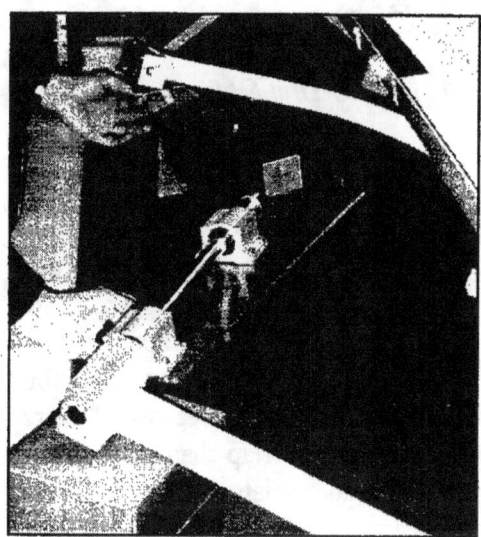

Figure 3. Laser Device for Checking Scattering Angle

Detailed documentation about the visibility sensor fork production history is presently not available to the government. Ideally, it would be desirable to know the fork serial numbers produced before and after the jig change and the fork serial numbers tested with (a) the laser device, (b) the first CMM program, and (c) the complete geometry CMM program. The available data from the complete CMM program will be discussed in Section 3.6.3. Fortunately, enough complete fork measurements were available to give reasonable confidence in the entire production run.

1.6 SCOPE OF REPORT

Many steps for validating the calibration of the US RVR forward scattermeter have been completed and were documented in a comprehensive report,[1] which covered US activities through August 1996. This section will summarize previous activities and highlight the questions that remained unanswered at that date. The goal of the validation effort is to answer *all* unanswered questions. Because questions remain after the efforts reported here, this report has been designated an interim report. The questions remaining and an assessment of their importance will be presented in the recommendations of Chapter 6.

Since 1994 the US and the UK have engaged in a joint RVR test program which includes testing in both the US and the UK. The main thrust of this test program will be reported separately. However, information relevant to the issues of this report also will be included here.

Since deployment approval in August 1994, the RVR Program has continued to monitor some airport installations to detect any system anomalies (such as the offset mentioned in 1.3.2.5) and to assess the influence of severe weather on sensor performance. The study of severe weather performance will be reported separately.

1.7 BACKGROUND AND ADDITIONAL WORK REQUIRED

1.7.1 Unit-to-Unit Calibration Consistency

The successful use of a scattering plate to transfer VS calibration from one forward-scatter unit to another depends upon the unit-to-unit consistency of the ratio of plate scattering to volume scattering from the obstruction to vision. The consistency of this ratio depends upon the manufacturing consistency of the scattering geometry from one unit to the next.

A fog calibration simulation model was developed by the Volpe Center to assess the effects of geometry errors on the resulting VS calibration. The calibration was found to be most sensitive to scattering angle changes (because of the rapid decrease in fog scattering crossection with scattering angle) and beam overlap changes. Making one beam larger than the other at the calibrator location decreases beam overlap effects at the cost of making the calibration sensitive to the beam size of the larger beam.

The simulation model was somewhat validated by measurements of the look-out version of the Teledyne VS. The model was used to guide the 1993 design of the look-down version. The receiver beam size is more consistent than the transmitter beam size; consequently, the calibration plate location was moved closer to the transmitter and away from the receiver so that the receiver beam would be larger at the calibrator location. The model also guided the production tolerances for the forks of the look-down version. The scattering angle was restricted to 42 ±0.25. The intercepts of the two beam centers at the calibrator were restricted to be no more that 0.30 inches apart.

Eventually the production tolerances of the national deployment VS were controlled by measuring each fork with a CMM. In the summer of 1995 the CMM program was modified to output all the fork geometry parameters needed to simulate the calibration. All forks subsequently produced were measured with this program, so that their calibration can be simulated. The simulation model was revised to use the CMM measurements as input. Earlier forks were characterized only by measurements of the production control tolerances; consequently, their calibrations cannot be simulated. (In June 1997, CMM measurements were made for the early production forks field tested in the US and the UK.)

Teledyne analyzed 87 forks and used the Volpe Center's simulation program SIMCAL8 (a) to assess compliance with the ± 7% variation permitted by the RVR specification (calculated variance was ± 5%) and (b) to select "golden" forks with close to correct calibrations to be used in field testing to determine a definitive fog calibration. The field testing of the selected forks gave fog calibrations consistent to ± 2% for some test periods. Because the field measurements include the effects of head geometry variations, which were not modeled, this consistency suggests that head variations are relatively unimportant (±2% or less) in defining calibration consistency (allowed variation of ±7 %).

The following questions remain to be answered concerning unit-to-unit calibration consistency:

1. How valid is the calibration simulation model?
2. How can the calibration be validated for forks not having complete CMM geometry?

These issues are addressed in Chapter 3.

1.7.2 Calibration Simulation Model Validation

Validation of the simulation model is important because it forms the basis for determining that the unit-to-unit variation in fog calibration is within the specified tolerances of ±7%. Three types of validation are needed.

1.7.2.1 Correct Computer Code

Is the computer code correct? Chapter 3 will address this question; in fact, two errors were found in the SIMCAL8 program, one of which had a significant impact on calibration accuracy.

1.7.2.2 Completeness

SIMCAL8 models only the effects of fork geometry on calibration; the heads are assumed to be perfect. The added effect of head variations will be addressed in Chapter 4.

1.7.2.3 Correct Scattering Assumptions

SIMCAL8 assumes that the scattering angle dependence of fog and plate scattering is fixed at nominal linearized values. Chapter 4 examines the correctness of these values and the validity of the linear assumption.

1.7.2.4 Field Test Validation

Field testing (Chapter 2) should yield scattermeter calibrations that are consistent with the calculated calibrations. Only two tested forks have calculated calibrations which deviate significantly from the golden forks. Unfortunately, predictions for these forks are not consistent with available measurements (see Chapter 4).

1.7.3 Calibration Determination

1.7.3.1 Calibration Methodology

Scattermeter errors were summarized in Table 1; errors can be classified as slope and offset.

Slope Errors — The slope calibration of the national deployment VS sensors is determined by comparisons with the transmissometers at the Otis Weather Test Facility, located on Cape Cod, Massachusetts. A number of sensors are calibrated with a master calibrator plate and then compared to the transmissometers to determine their exact fog calibrations using the box-plot method. Note that this calibration process is probably consistent to 1 or 2%.

Offset Errors — The final design of the national deployment sensor heads was modified (from the "new Denver" version) to minimize scattered light by increasing the size of the hoods and making all possible scattering surfaces black. (Unfortunately, some national deployment sensors still have significant electronic offsets that are large enough to affect sensor RVR accuracy. A sensor modification correcting this problem is currently under evaluation, but will not be considered in this report.)

1.7.3.2 Calibration History

Three national deployment VS units were installed at Otis in July 1994. Their initial measurements were used to provide an interim calibration for national deployment. More complete data from Otis and two other sites (Portland, OR, and Birmingham, England) were analyzed in January 1995; the results suggested that the interim calibration be corrected by approximately -3.5%. Because this correction was small and more representative data was expected shortly, the decision was made to keep the interim calibration until a more definitive result was obtained.

In January 1995 the decision was made to send five golden VS units to Otis for calibration testing. The golden sensors were shipped to Otis and installed in early January 1996. Measurements on the golden-fork VS through August 1996 suggested that the calibration needed to be changed by +6%. (Data collected through June 1999 confirmed this result.)

Chapter 2 will extend the calibration history through the date of this report and present data analyzed using a different method for defining test periods.

1.7.4 Calibration Transfer

The final calibration transfer station consists of two VS units with sensor interface electronics (SIE) and a single data processing unit (DPU). The VS units are calibrated with the master calibrator. A new calibrator then is measured for 3 minutes in one VS and then checked for consistency in the other VS. The 3-minute value is inscribed on the label. The process is repeated if the two readings disagree by more than a defined amount. For consistency and convenience, the calibration transfer process is controlled by a personal computer that receives data from the DPU's engineering data port (EDP).

1.7.4.1 Original Procedure

The first few calibrators used for national deployment were measured in the three Otis VS units. The master Otis calibrator then was sent to Teledyne for routine calibration of subsequent calibrators.

1.7.4.2 Interim Procedure

Until the RVR system is taken over by the FAA, the calibration transfer station remains at Teledyne. Two golden forks and heads were selected for use. Two master calibration plates were sent to Otis for calibration using the original procedure.

1.7.4.3 Long-Term Procedure

For the long term, the calibration transfer station will be moved to FAA logistics branch (AML) in Oklahoma City, OK.

1.7.5 Calibration Accuracy

Teledyne provided all available complete CMM data to the Volpe Center. Chapter 3 describes the analysis of this data.

1.7.6 Long-Term Calibration Maintenance

If the VS calibration is to be assured over the life of the system, the system calibration must be periodically referenced to transmissometer measurements. The frequency and nature of this recalibration will depend upon the observed stability of the VS calibration. Calibration drift can occur from changes in either the calibrator (e.g., deterioration of the plastic) or the VS (e.g., relaxation of the fork bends or asymmetrical interior contamination of the head windows). Two approaches can be used to assess calibration stability:

1. Direct assessment of stability: (a) sensor/calibrator examination and remeasurement, (b) accelerated aging tests, and (c) comparisons with sensors/calibrators stored under optimum conditions.

2. Repetition of original calibration procedures for referencing airport VS calibrations to transmissometers.

An RVR system and the reference transmissometers should be operated at the Otis WTF for the life of the system. After Air Force operation of the site was terminated, it was taken over by the Department of Transportation. The site will be kept clear of trees. If other site uses no longer require on-site personnel, the site will be operated as a remote site.

Consistency checking will begin after the FAA has taken over the RVR system and has the calibrator marking station and the CMM machine. The first step will involve remeasuring a random sample of calibrators and forks. Eventually, all calibrators will have to be recycled for the 6% calibration change. Periodic remeasurements (2 to 5 years) for calibrators and forks will be instituted; the frequency of checks will be based on the initial random tests.

Because the fork geometry can be damaged in ways that are not obvious (e.g., when a falling pole hits a stop abruptly in such a way that the sensor heads do not hit the ground, but the fork might be distorted by inertial forces), it might be desirable to fabricate a template that can check fork geometry in the field.

2. FIELD CALIBRATION

This chapter presents the test results for the final design of the US RVR forward scattermeter. Most of the data comes from tests[1] at the Otis Weather Test Facility, located on Cape Cod, Massachusetts. Some data are presented from tests[6] at the Birmingham, England, Airport.

2.1 TWO RVR SYSTEMS

The forward scattermeter consists of three components: transmitter head, receiver head and mounting fork, all of which were manufactured by Handar, Inc. The prime RVR contractor, Teledyne Controls. Inc., designed the SIE and processing software. The national deployment SIE was based on 1988 technology and has hardware, computer memory and speed, and software limitations for developing and testing improvements to the original design. In 1995, Teledyne developed a new SIE based on PC-based hardware and software which was used to evaluate preplanned product improvements (P^3I) to the original design. The same scattermeter hardware can be used with either SIE. Because the basic scattermeter performance is defined by the sensor heads and scattering geometry, one would expect the sensors to have the same calibration characteristics with either SIE. In the following presentations the national deployment SIE will be indicated by the parameter name TDNn, n=1,6 and the P^3I SIE by the parameter name TVSn, n=1,3.

2.2 EARLY PRODUCTION UNITS AT OTIS

The decision to deploy the new generation RVR system was made in August 1994. Deployment required that the sensor calibration be defined. The preliminary sensor calibration used the first

Table 3. Preliminary Calibration Sensors

Serial Numbers

Name	Fork	Rx Head	Tx Head	Min. Data	CMM RVP.	CMM RPV	RVP +1.25	RPV +1.25
TDN1	407	714	734	455	1.028	0.973	1.037	0.964
TDN2	360	636	622	107	1.069	0.935	1.032	0.969
TDN5	389	694	715	84	1.005	0.995	1.209	0.829

national deployment sensors that became available. The first three units were early production units where the forks were bent to meet the scattering angle specification (see Figure 1 for calibration fixture). These forks were measured on June 12, 1997, long after they were field tested; the calculated calibration results are listed in the last columns of Table 3.

2.2.1 Relationship of Field Test Results to Calibration Simulation Results

The field test results define the scattermeter fog calibration as the median ratio of scattermeter MOR to transmissometer MOR. The calibration simulation calculates RVP, that is, the ratio of volume scattering (i.e., fog) to plate scattering (i.e., calibrator). A scattermeter is calibrated by setting its reading of the calibrator scattering to the nominal extinction coefficient of the calibrator. If a scattermeter's RVP value is greater than 1, then it will measure a fog extinction coefficient that is greater than the actual extinction coefficient. In this case, the fog MOR will be less than the true MOR and the median MOR ratio will be less than 1. Thus, the median fog MOR should be proportional to RPV, the ratio of plate scattering to volume scattering (RPV = 1/RVP). When available, the RVP and RPV values will be listed for the scattermeters tested. Note that all values in this chapter are from the corrected simulation program, SIMCAL9, using the Case 3 parameters of Section 4.2.3.

2.2.2 Preliminary Fog Calibration

In June 1994, the three national deployment forward scattermeters listed in Table 3 were installed at Otis. Data from the fog events in June were used to derive a preliminary calibration for the sensors. The minutes[†] of data listed in Table 3 are different for the three sensors because they were installed at different times. Because the three scattermeters were calibrated using S/N 0001 with the incorrect plate location (see Section 4.3), the volume-plate scattering ratios also were calculated for a calibrator location of +1.25 inches.

The master calibrator (S/N 22) was measured in the three sensors listed in Table 3. Its nominal calibration value was adjusted (final value = 33.5 /km) to give the best fit to the fog calibrations of the three sensors with respect to the average of the crossed visible light reference transmissometers (Section 5.2.2.1). The first calibrators used at airports then were measured in the same three sensors after they had been calibrated by the master calibrator.

2.2.3 Preliminary Calibration Check

The preliminary calibration was checked in January 1995 by analyzing additional data (period 7/16/94-12/21/94) from the same three sensors listed in Table 3. The results are presented in Table 4. Although the longer test period suggested a possible change in calibration of approximately 3%, no changes were made until the golden fork sensors could be tested. Testing of the three sensors continued through December 1995.

Table 4. Preliminary Calibration Check

Sensor	Minutes	Median MOR Ratio
TDN1	1244	1.050
TDN2	1315	1.009
TDN5	938	1.038

2.3 GOLDEN UNITS AT OTIS

In January 1996, five golden scattermeters were installed at Otis. The selection of the five golden sensors (TDN1 through TDN5) was described previously[1] (Section 5.1.3.2). They were selected to have nominal fork scattering geometry and have calculated calibrations near the middle of the calculated calibration distribution. Table 5 lists the seven Teledyne visibility sensors that were installed at Otis during the test periods. The five national deployment baseline (NDB) units

Table 5. Golden Scattermeters at Otis

Sensor Name		Serial Numbers						
NDB	P³I	Fork	Rx Head	Tx Head	CMM RVP	CMM RPV	RVP +1.25	RPV +1.25
TDN1		287	455	295	1.002	0.998	1.245	0.803
TDN2	TVS3	107	262	202	1.005	0.995	1.228	0.814
TDN3		98	122	158	0.999	1.001	1.230	0.813
					1.011	0.989	1.255	0.797
TDN4		201	297	257	0.999	1.001	1.250	0.800
					1.014	0.986	1.262	0.792
TDN5		481	865	906	0.992	1.008	1.195	0.792
					0.993	1.007	1.258	0.795
	TVS1	365	486	548	N/A			
	TVS2	224	333	275	1.013	0.987	1.224	0.817

[†] The fog minutes are for extinction coefficients above 5 /km and homogeneity of better than 10%, according to the US RVR specification.

are termed TDN1 through TDN5. The three P^3I units are termed TVS1 through TVS3. One sensor (fork S/N 107) was switched from NDB to P^3I during the first test period. The calibrations calculated from CMM data are listed in the last column. Only TVS1 had no fork measurement. The measurements for TVS2 showed that it could be classified as golden. Two measurements were made for some forks. The calculated calibrations are equal to 1.00 ±0.02 relative to the mean calibration of 87 units. All the scattermeters at Otis were calibrated with calibration plate S/N 22.

2.4 PRELIMINARY UNITS AT BIRMINGHAM

In December 1994, two early production national deployment scattermeters were installed at the Birmingham, England, Airport. Table 6 lists the information about these sensors.

Table 6. Scattermeters at Birmingham (1994-95)

Name	Serial Numbers				
	Fork	CMM RVP	CMM RPV	RVP +1.25	RPV +1.25
TDN1*	515	1.014	0.986	1.103	0.907
TDN2*	520	1.001	0.999	1.185	0.844

* Identification of fork and name is lost.

2.5 GOLDEN UNITS AT BIRMINGHAM

In November 1995, the national deployment scattermeters at Birmingham were replaced by the two golden P^3I scattermeters listed in Table 7. After completion of the Birmingham testing, these units were returned to Otis and tested using national deployment SIEs.

Table 7. Scattermeters at Birmingham (1995-97) and Otis (1998)

Sensor Name		Serial Numbers				
Birm.	Otis	Fork	Rx Head	Tx Head	CMM RVP	CMM RPV
TVS1	TDN2	175	526	471	1.003	0.997
TVS2	TDN6	227	658	576	1.000/1.000	1.000/1.000

2.6 MEDIAN FOG CALIBRATIONS

2.6.1 First Analysis

The first analysis was conducted as the data was being collected. The test periods were assigned arbitrarily according to when blocks of data became available.

Figure 4 shows the first analysis results for the first two years of testing for the final US RVR scattermeter design. The average of the crossed transmissometers (termed "TAVE") was used as reference and a 10% homogeneity criterion was applied. Data were included for

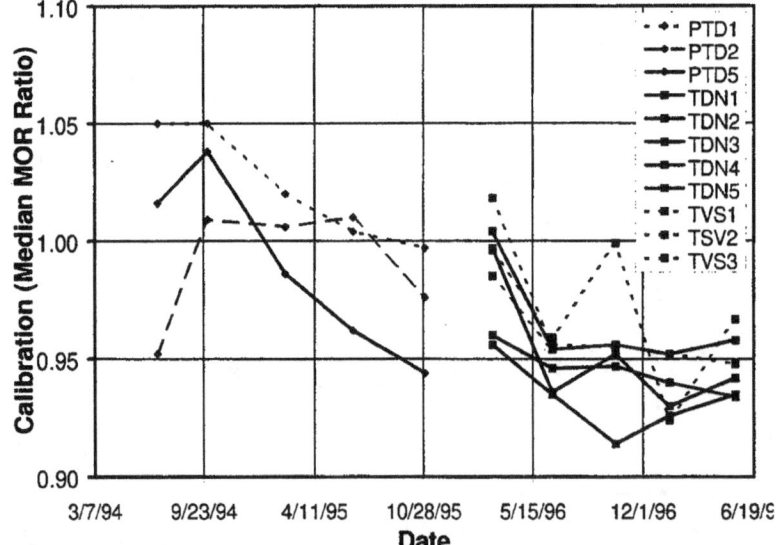

Figure 4. Otis Fog Calibration Results (1994-97)

reference $\sigma > 5$ /km (MOR < 650 m).

The break in the middle of Figure 4 separates the preliminary calibration sensors from the golden sensors. The results in Figure 4 show some of the predictions of the CMM calculations, but also show some variations that are difficult to understand.

1. The spread in scattermeter fog calibration is very small (better than ± 3%) for the 2nd, 4th, and 5th golden sensor test periods. This consistency suggests that the selection criteria for golden sensors actually results in consistent fog calibrations. The typical median MOR ratio for these periods would be 0.94 or 0.95. At the end of this analysis it appeared that instead of the +3.5% calibration error suggested by the preliminary calibration check, the actual calibration error was -5.5%.

2. The 1st golden sensor test period shows higher median MOR ratios and greater unit-to-unit spread than the later periods. Although this shift might be related to different fog characteristics, it also might reflect poor performance of the reference transmissometers during this period. The transmissometers often were affected by blowing snow; the snow effects might not have been completely filtered out of the analysis by the 10% homogeneity criterion and hence might have biased the reference data.

3. The relative calibration of the national deployment SIE golden sensors (TDNn) remains somewhat consistent over the five test periods. This observation is consistent with the concept that the sensor calibration is a function of sensor geometry which remained fixed over the test periods. The P^3I sensor showed greater period-to-period calibration variation than the national deployment sensors.

4. The preliminary test periods show considerable scatter, particularly for the earliest test periods. The last two preliminary test periods show results more consistent with the CMM calculations. TDN5 has a median fog calibration close to the typical golden values while TDN1 and TDN2 have a median fog calibration significantly higher.

5. TDN5 gives evidence for a time varying fog calibration. A possible explanation is that the bent fork relaxed to its original configuration with age.

The first-analysis results for Birmingham will not be included.

2.6.2 Second Analysis

The second analysis was conducted long after the data were collected and selected test periods were based on the actual occurrence of fog. The goal was to group fog periods lasting 2 to 4 weeks into separate analysis files. Data were not included from periods with mostly inhomogeneous data since such inhomogeneity may reflect transmissometer performance problems that may bias the data.

Figure 5 shows the second analysis of the Otis results over the period 1994-98. In contrast to Figure 4, no P^3I sensors are included and the preliminary sensors are plotted with the same names as the golden sensors. In Figure 5 the gap in the data lies in the middle of the preliminary sensor data, not between the preliminary and golden data. The boundary between the preliminary and golden data is the vertical tick at 12/29/95. Note that the tick marks in Figure 5 are set for quarter years.

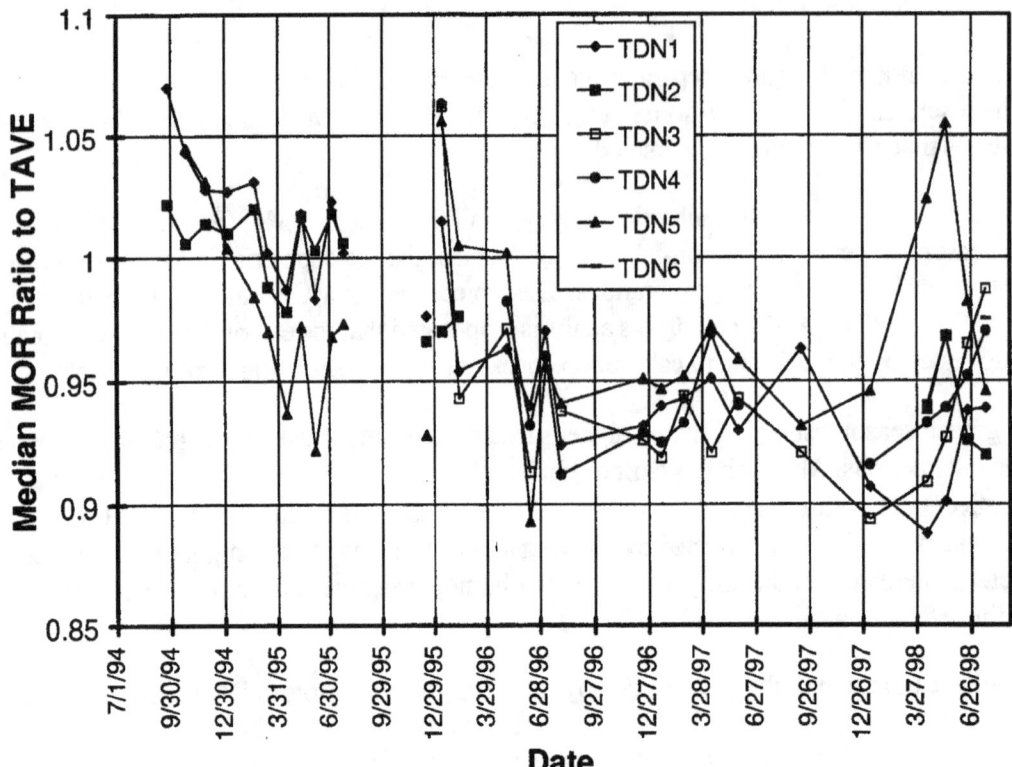

Figure 5. Fog Calibration of Otis "Deployed" Scattermeters

The following comments can be made about the results in Figure 5:

1. Although the time resolution is finer, the results generally agree with Figure 4.

2. Most of the anomalous calibration for the first golden test period occurred in the first month of 1996.

3. The sensor calibration drifted during the second quarter of 1998. Calibration checks showed that the TDN2 was correct, but the TDN1 calibration was 4% low and the TND5 calibration was 10% high. These calibration errors explain the deviations of these two sensors from the nominal 0.94 fog MOR ratio for that time period.

The relative calibration of scattermeters can be determined by using one scattermeter to evaluate the others. Figure 6 shows the preliminary sensor data where TDN2 was used as the reference sensor. Because the performance of the reference transmissometers is not an issue, no homogeneity criterion was applied. Consequently, reasonably reliable median results can be obtained for all test periods. In Figure 6 the periods not plotted in Figure 5 are enclosed with boxes.

Figure 7 plots four years of Birmingham test results. The first data (1994-95) is for the sensors listed in Table 6. After the break in the lines, the results are plotted for the sensors in Table 7. The sensors used P^3I SIEs at Birmingham for 1995-97 and national deployment SIEs at Otis for 1998.

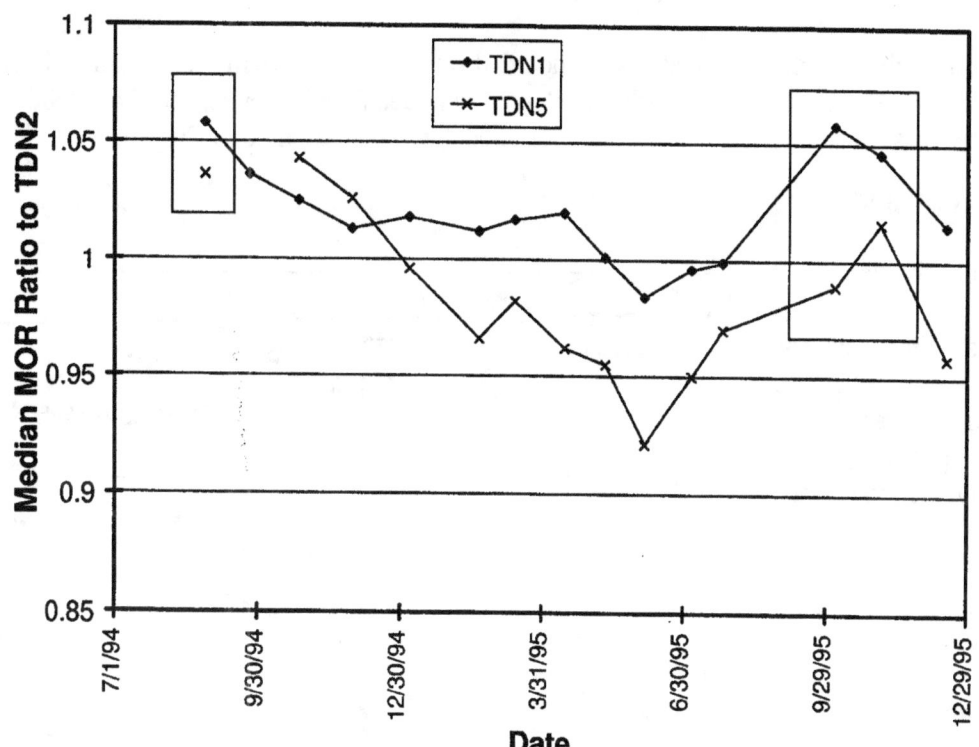

Figure 6. Intercomparison of Early Production Deployed Scattermeters

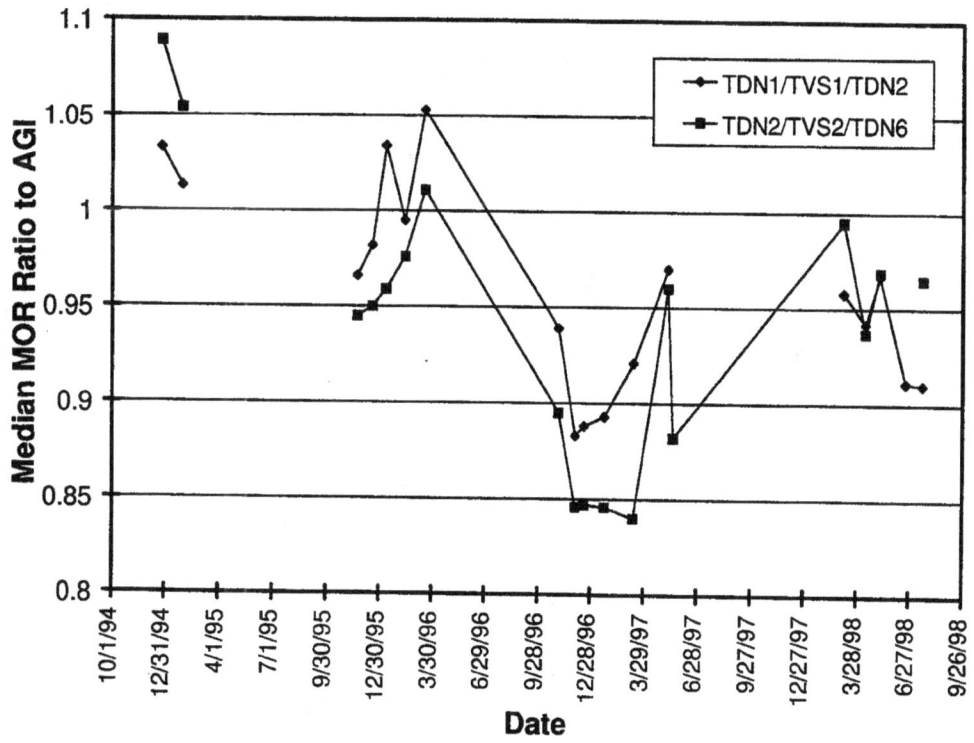

Figure 7. AGI Calibration of Birmingham Scattermeters

3. CALIBRATION SIMULATION

This chapter describes the standard analysis method that was used to derive the calibration validation results presented in Reference 1. This method is described in detail and extended to all available CMM data. Extensions to the calibration simulation model will be presented in Chapter 4.

3.1 FORK GEOMETRY

The original calibration simulation model used an idealized scattering geometry that would be very difficult to convert to the Handar measurement geometry. Subsequently, the model was rewritten in the Handar drawing geometry (see Figure 8); this version was called SIMCAL7. Unfortunately, not all the needed parameters were recorded by the CMM. In July 1995, the CMM program was modified to output and store the needed information. The corresponding simulation program is called SIMCAL8.

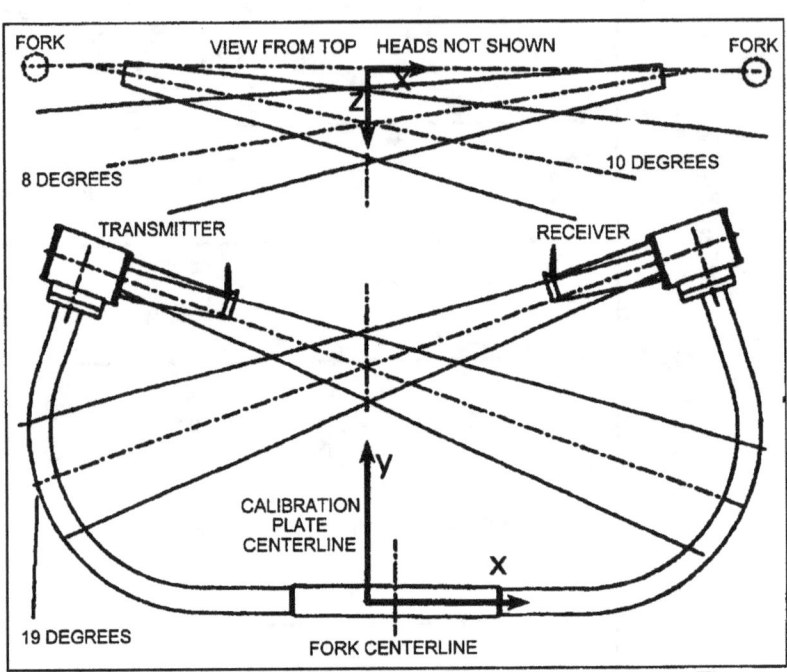

Figure 8. Scattermeter Drawing: Top and Side Views That Define SIMCAL7 Coordinate System

Figure 9 shows photographs corresponding to Figure 8. The calibrator plate is mounted on the transmitter side of the mounting post.

Figure 10 shows the black side of the calibrator plate and its mounting bracket. The mounting bracket screws into the back side of the fork as viewed in the side views of Figures 8 and 9.

Figure 11 shows the scattermeter offset from the plane view in Figure 9. The black side of the calibrator plate points toward the transmitter and the white (diffuser) side of the plate points toward the receiver.

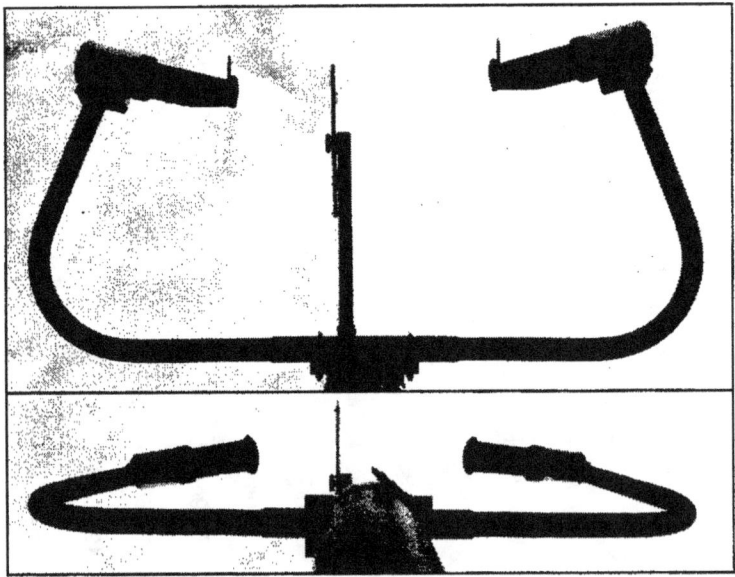

Figure 9. Scattermeter Photographs (side, bottom)

Figure 10. Calibration Plate

The drawing geometry (Figure 8) used in SIMCAL7 references the head positions to the bottom of the fork. Unfortunately, the CMM cannot measure the bottom of the fork which defined the SIMCAL7 x axis. Consequently, SIMCAL8 uses a different geometry that is

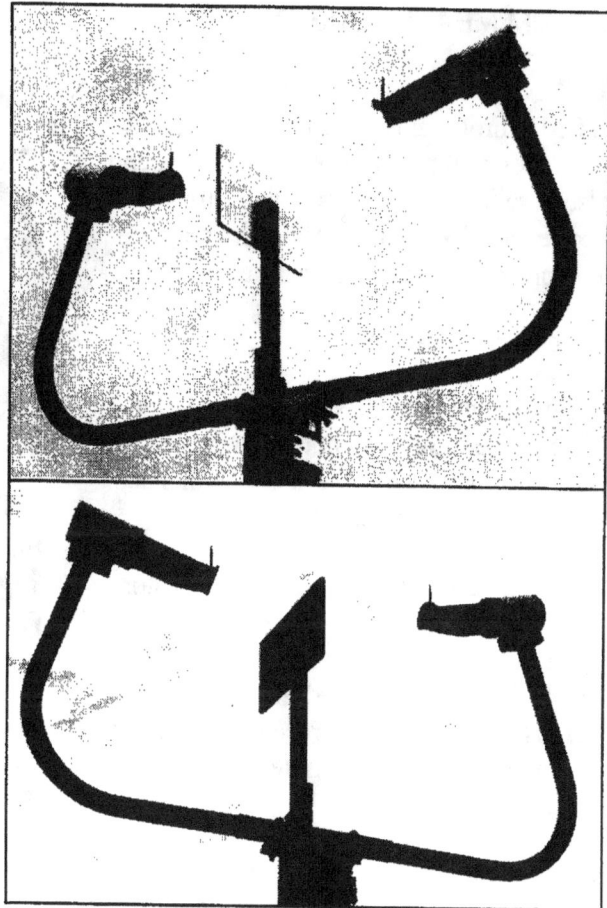

Figure 11. Offset Views of Scattermeter with Calibrator Installed

closely related to the drawing geometry: (a) the origin of the x and y axis is different and (b) the three axes are defined by their relationship to the calibrator plane and transmitter mounting plate plane rather than the plane of the fork.

The calibrator plate location is simulated by a fixture that is mounted on the calibrator mounting bracket and provides a measurement plane that is approximately equivalent to that of the calibration plate. The measurement plane is actually the plane of the calibrator mounting bracket. As seen in Figure 9, this plane is slightly displaced (by +0.13 inches) from the black attenuation mask of the calibration plate. The measured plane is corrected by -0.13 inches and then used to define the y-z plane and the zero for the x axis. The z axis is defined by the intersection of the plane of the transmitter mount and the calibrator plane. The origin of the y-z plane is taken as the intercept at the calibrator of the line between the center of the tubing hole in the receiver mount and the center of the tubing hole in the transmitter mount. The CMM measurements used in the simulation are referenced to this coordinate system.

3.2 CMM MEASUREMENT

The CMM machine is Mitutoyo Model BHN 715. The CMM program is presented in Appendix A; its operation will be summarized here. Appendix A also describes the formats of the hard copy and stored reports. The process of determining the fork geometry will be described step by step.

First, the fork is installed:

1. The technician mounts the fork (see Figure 12) on the measurement table. A three-point mount is used and the fork is held in place with a vacuum system.

2. The technician then mounts the calibrator plate fixture to the fork (see Figure 13). Note that, if the fork were viewed from above the table, it would be viewed from the back side of Figures 8 and 9. The receiver mount is the far mount and the transmitter mount is the near mount.

The measurements start in the native coordinate system of the CMM. The first measurements are made on the receiver mount, which is more closely fixed to the CMM table than the transmitter mount (see Figure 14). A close view of the receiver head is shown in Figure 15.

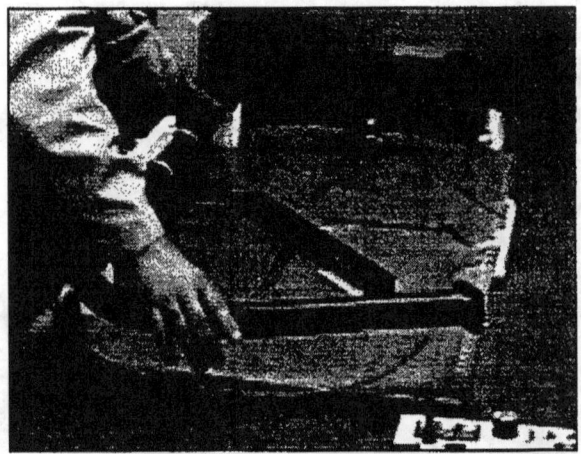

Figure 12. Installing Fork

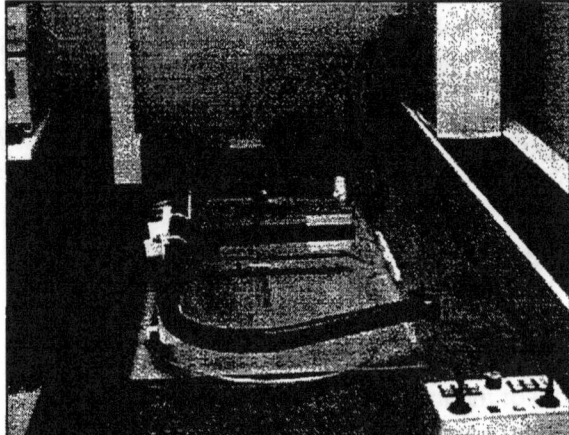

Figure 13. Fork in Place with Calibrator Plate Fixture Installed

Figure 14. Receiver Mount Measurements

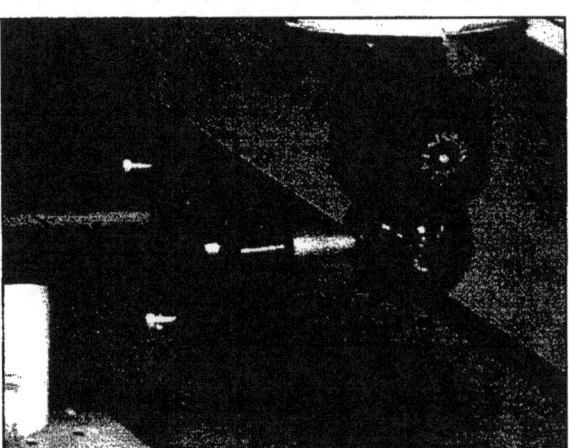

Figure 15. Receiver Head, Close View

The following steps are taken to measure the receiver head mount:

1. Determine the plane (three points) of the top of the receiver head mount.

2. Measure two sides of the receiver mounting plate.

3. Remeasure plane and flatness (eight points) of receiver head mount. The orientation of the normal to the plane (away from the fork) is specified by two angles Θ_r (the angle with respect to the x-axis) and Φ_r (the angle with respect to the y-axis in the y-z plane, positive for positive y and positive z).

4. Measure the locations of the alignment pins in the receiver head mount. Use the nominal pin locations on the mounting plate to calculate the vector of the nominal receiver beam direction in the plane of the receiver head mounting plate. Determine position of nominal center of large hole in receiver mounting plate (through which beam vector passes).

Figure 16. Manual Setting for Transmitter Mount

The following steps are taken to measure the transmitter mount:

1. Determine the plane (three points) of the transmitter head mount (see Figure 16). This measurement is done manually to accommodate variations in transmitter mount locations which are too great for automatic measurement.

2. Measure two sides of the transmitter mounting plate.

3. Remeasure plane and flatness (eight points) of transmitter head mount. The orientation of the normal to the plane is specified by two angles Θ_t (the angle with respect to the x-axis) and Φ_t (the angle with respect to the y-axis in the y-z plane, positive for positive y and positive z).

4. Measure the locations of the alignment pins in the transmitter head mount. Use the nominal pin locations on the mounting plate to calculate the vector of the nominal transmitter beam direction in the plane of the transmitter head mounting plate. Determine position of nominal center of hole in transmitter mounting plate (through which beam vector passes).

5. Determine intersection of the planes of the two mounting plates. This intersection is approximately perpendicular to the plane of the fork. This intersection then is used to calculate the in-plane and out-of-plane beam angles.

6. The calibration plate plane then is measured: Measure the plane (four points) of the calibrator (see Figure 17). This plane is the y-z plane of the calculations. The x-axis is perpendicular to the

calibrator; a nominal displacement of the calibrator from the mounted measurement fixture is made (-0.13 inches).

7. The distance then is calculated between the intersections of the transmitter and receiver beams with the calibrator plate (nominal beam locations are in the planes of head mounting plates and are not the actual location in heads, which are offset from mounting plates, as shown in Figure 18).

Figure 17. Calibration Plate Fixture Measurement

The coordinate system then is changed to match the coordinate system of the calculation:

1. The y-z plane is defined as the calibrator fixture displaced by -0.13 inches.

2. The z-axis direction is defined by the intersection of the y-z plane with the plane of the transmitter mounting plate.

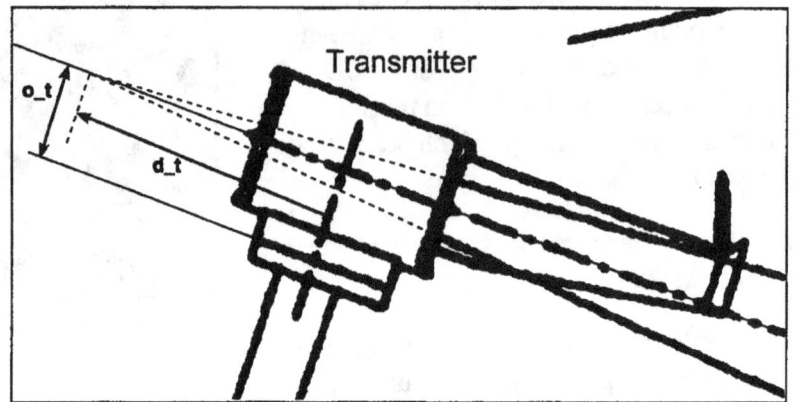

Figure 18. Transmitter Head Detail

3. The origin of the y-z plane is defined as the intersection of the line between the receiver mount hole nominal center and the transmitter mount hole nominal center.

The six parameters needed to define the calculation are then output:

1. Receiver mount hole center (x_r, y_r, z_r).

2. Transmitter mount hole center (x_t, y_t, z_t).

3. Intersection (yc_r, zc_r) of receiver beam (in plane of mounting plate) with y-z plane.

4. Intersection (yc_t, zc_t) of transmitter beam (in plane of mounting plate) with y-z plane.

5. Plane of receiver mounting plate (normal given by Θ_r, Φ_r).

6. Plane of transmitter mounting plate (normal given by Θ_t, Φ_t). Φ_t is zero by the definition of the z axis of the coordinate system.

3.3 HEAD GEOMETRY

In addition to the fork measurements, the nominal head geometry must be included in the calculation. The geometry of the transmitter head is shown in Figure 18. It is assumed that the receiver and transmitter beams are centered on and parallel to mounting plate with offsets of o_r and o_t (both 2.44 inches), respectively. The effective beam origins are displaced amounts d_r and d_t (4.69 and 7.18 inches), respectively, from the hole center position in the plane of the head mount. With these new effective beam origins, the nominal beam width is reduced compared to the values used in earlier calculations, which assumed that the beam origin was at the head. The resulting angle widths (half angle, half response) are 4.07 and 4.28 degrees for transmitter and receiver, respectively. The previous beam taper values (40% for receiver, 25% for transmitter) are used. At present all these head parameters are kept fixed in the calculation. In Chapter 4 the effects of head orientation errors will be introduced by varying the two intercepts of the beams with the calibrator.

3.4 SIMCAL8 PROGRAM

The goal of the calibration simulation is to calculate the ratio between volume scattering from fog and scattering from the calibration plate. This ratio determines the validity of the use of a calibrator plate to calibrate a visibility sensor.

3.4.1 Fixed Parameters

The fixed calibration simulation parameters are read once from a file called CAL.DAT. In addition to the head parameters mentioned in Section 3.3, the calibration simulation uses the following assumptions and parameters:

The scattering is calculated from a uniform grid of points:

    vdelta = the grid increment for the volume calculation

    delta = the grid increment for the plate calculation

The values vdelta = 0.15" and delta = 0.03" were found to give reasonable stable scattering values (17,000 plate points, 30,000 volume points).

The scattering from the volume and plate are assumed to vary linearly about a nominal scattering angle for the sensor:

    theta = the nominal angle for the differential scattering crossection

    b_vol = the fractional change in volume scattering crossection for 1 degree change in scattering angle (now assumed = -0.07)

    b_cal = the fractional change in plate scattering crossection for 1 degree change in scattering angle (now assumed = -0.03)

A distance scale d is used in the calculations to give reasonable scattering levels (e.g., 1 and 20)

A number of parameters are used to characterize the receiver and transmitter beams, which are taken as cylindrically symmetric in SIMCAL8 (the transmitter asymmetry parameter ft might be reactivated in the future):

rz	=	angle from transmitter beam center to half response
tz	=	angle from receiver beam center to half response
pr_edge	=	full width of receiver beam edge taper as percent of rz

Note that a percentage taper of 50% for pr_edge gives full response out to 0.75 rz and zero response at 1.25 rz.

pt_edge	=	full width of transmitter beam edge taper as percent of tz
ft	=	defines a y-axis asymmetry in the transmitter beam (not used in SIMCAL8 at present)

The response at the top of the beam at angle tz is increased to (1+ft) tz and at the bottom of the beam at angle tz reduced to (1-ft) tz.

The standard SIMCAL8 contents of CAL.DAT are:

vdelta = 0.15"; delta = 0.03"; theta = 42.0°; b_vol = -0.07/deg; b_cal = -0.03/deg

d_t = 7.18"; d_r = 4.69"; o_t = 2.44"; o_r = 2.44"; d = 22.0"

tz = 4.07°; rz = 4.28°; pt_edge = 25.0%; pr_edge = 46.0%

Table 8 lists the half-angle shapes of the transmitter and receiver beams based on these parameters. The receiver has larger half and zero response sizes, but a smaller full response size than the transmitter.

Table 8. Beam Size (mrad)

Beam	Full	Half	Zero
Transmitter	62	71	80
Receiver	50	75	92

3.4.2 CMM Parameters

The variable parameters of the simulation are read from a file called SIMCAL8.DAT. These parameters are extracted from the data files stored by the CMM machine by using a macro with the WordPerfect ASCII Editor: ED.EXE. Note that any blanks in the fields or blank fields must be filled to give a file that can be read correctly by SIMCAL8 (e.g., blanks in the date or missing serial numbers). Four forks were measured on 7/21/95 and the extracted data values are listed in Table 9.

The output of each case is sent to one line in the file SIMCAL8.PRN in comma delimited format, suitable for importing into a spreadsheet. Table 10 presents the outputs for these four cases using the standard CAL.DAT file parameters above. Some of the results also are shown at the bottom of the table for the smaller values of vdelta = 0.05" and delta = 0.01". The standard values for vdelta and delta lead to approximately 30,000 and 18,000 scattering points for volume and plate scattering, respectively. The smaller values lead to 160,000 and 840,000 points, respectively. The results of the simulation are virtually unchanged by the increased number of points; thus, the standard values can be expected to include enough scattering points to yield stable results.

Table 9. Sample Input Data for SIMCAL8

Date of Measurement	7/21/95	7/21/95	7/21/95	7/21/95
Assembly S/N	234	182	137	134
Receiver Mount Hole X, x_r	23.0732	23.2625	23.1222	23.2049
Receiver Mount Hole Y, y_r	1.0252	0.7109	0.9817	0.8555
Receiver Mount Hole Z, z_r	0.3348	0.4130	0.4133	0.4489
Transmitter Mount Hole X, x_t	-19.5552	-19.3551	-19.5484	-19.4249
Transmitter Mount Hole Y, y_t	-0.8689	-0.5915	-0.8300	-0.7161
Transmitter Mount Hole Z, z_t	-0.2838	-0.3437	-0.3494	-0.3758
Receiver Beam Intercept Y, yc_r	-7.2325	-7.2042	-7.2492	-7.2416
Receiver Beam Intercept Z, zc_r	3.5957	3.4760	3.4665	3.5781
Transmitter Beam Intercept Y, yc_t	-7.3782	-7.2765	-7.3368	-7.2731
Transmitter Beam Intercept Z, zc_t	3.5887	3.6851	3.5464	3.7189
Receiver Mount Plane Angle, Φ_r	0:11:48	-0:00:17	-0:11:21	-0:27:56
Receiver Mount Plane Angle, Θ_r	109:40:02	108:47:29	109:36:59	109:17:28
Transmitter Mount Plane Angle, Θ_t	71:36:13	70:56:14	71:35:32	71:24:15

Table 10. Sample Output Data for SIMCAL8

Date of Measurement	7/21/95	7/21/95	7/21/95	7/21/95
Assembly S/N	234	182	137	134
Number of plate scattering points	17590	17367	17704	17788
Number of volume scattering points	30910	30870	31143	30937
Mean plate scattering angle	42.09	41.86	41.82	42.01
Mean volume scattering angle	41.23	41.01	40.97	41.17
Standard deviation plate scattering angle	3.59	3.57	3.60	3.63
Standard deviation volume scattering angle	3.07	3.07	3.06	3.07
Plate scattering signal	1.6889	1.6907	1.7174	1.7236
Volume scattering signal	10.5295	10.7604	10.8152	10.6196
Ratio volume scattering to plate scattering = RVP	6.2347	6.3643	6.2974	6.1612
Number of plate scattering points	158344	156193	159330	160134
Number of volume scattering points	834078	833263	840323	836003
Plate scattering signal	1.6888	1.6907	1.7174	1.7235
Volume scattering signal	10.5283	10.7595	10.8142	10.6184
Ratio volume scattering to plate scattering = RVP	6.2341	6.3639	6.2969	6.1608

3.5 87 PRODUCTION UNITS

Teledyne Controls used SIMCAL8 to analyze the results of complete CMM measurements for 87 forks. The mean RVP value was 6.21 and was used to normalize the calibrations to that of a typical fork. Figure 19 (also presented previously[1]) shows the distribution of calculated calibrations for these 87 forks. The distribution is well within the ± 7% specification on unit-to-unit. Note, however, that the effects of

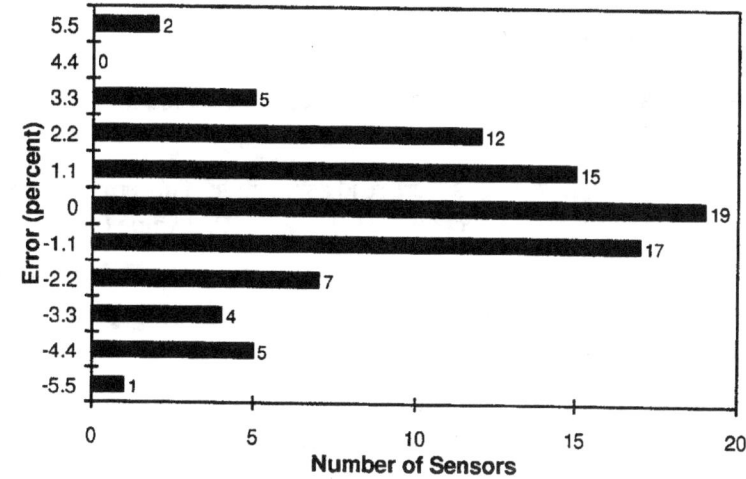

Figure 19. Distribution of Calculated Calibrations

head variations have not been included.

3.6 SIMCAL9 PROGRAM

As part of the calibration validation process, Teledyne provided all the available complete CMM measurement files for analysis. Because the forks tested before July 1995 used only the standard production quality control parameters, it was hoped that an analysis of these parameters would help validate the parts of the production run that lacked complete geometry information. Consequently, the quality control parameters were extracted from the data files along with the geometric parameters. The addition of these parameters and the method of data extraction required changes to the data input and output formats for the calibration simulation program. The new program was called SIMCAL9. The following quality control parameters (distances in inches) are taken from a measurement data file. Note that the long fork is the receiver and the short fork is the transmitter.

```
                actual     nominal    up_lim    low_lim    deviation
==================================================================
+++ FLATNESS OF PLANE ON LONG FORK. +++
FLAT      0.0048                0.0100
+++ DISTANCE BETWEEN PINS ON LONG FORK. +++
DST       3.5865     3.5850     0.0060    -0.0060     0.0015
+++ FLATNESS OF PLANE ON SHORT FORK. +++
FLAT      0.0055                0.0100
+++ DISTANCE BETWEEN PINS ON SHORT FORK. +++
DST       3.5864     3.5850     0.0060    -0.0060     0.0014
+++ COMPOUND ANGLE BETWEEN PLATES. +++
ANG     42:04:33   42:00:00    0:15:00   -0:15:00    0:04:33
+++ PROJECTED ANGLE BETWEEN THE PLATES. +++
ANG     38:04:43   38:00:00    0:30:00   -0:30:00    0:04:43
+++ THIS IS THE DISTANCE BETWEEN THE INTERSECTION POINTS. +++
DST       0.1567      0.000     0.3000     0.0000     0.1567
+++ PROJECTED ANGLE BETWEEN THE LONG FORK AND THE CENTERLINE. +++
PHI      8:47:26    8:42:00    1:00:00   -1:00:00    0:05:26
+++ PROJECTED ANGLE BETWEEN THE SHORT FORK AND THE CENTERLINE. +++
PHI    -10:35:07  -10:30:00    1:00:00   -1:00:00   -0:05:07
```

3.6.1 Error Correction

During the validation process, two errors were discovered in the SIMCAL8 code and were corrected in the SIMCAL9 code:

1. Small negative angles were read as positive. This error affected only the parameter Φ_r and had only a small effect on the calculated RVP value (usually less than 0.5%).

2. The calibrator plate was put at an x position of about 0.8 inches rather than the correct 0.0 inches. This error affected only the plate scattering, but had a significant effect on the resulting RVP values, increasing the variance.

Table 11 shows the changes in the results of the calculation and should be compared to the upper portion of Table 10. Only the plate scattering shows significant changes.

Table 11. Sample Output Data for SIMCAL9, Same Cases as Table 10

File Name	FILE1001	FILE1002	FILE1003	FILE1004
Number of plate scattering points	17955	17878	17948	17919
Number of volume scattering points	30910	30870	31141	30921
Mean plate scattering angle	41.84	41.61	41.57	41.74
Mean volume scattering angle	41.23	41.01	40.97	41.17
Standard deviation plate scattering angle	3.88	3.88	3.88	3.89
Standard deviation volume scattering angle	3.07	3.07	3.07	3.06
Plate scattering signal	1.8796	1.8947	1.9018	1.8925
Volume scattering signal	10.5295	10.7602	10.8154	10.6087
Ratio volume scattering to plate scattering = RVP	5.602	5.6791	5.6869	5.6056

Table 12. Statistics of SIMCAL9 Calculation for 87 Teledyne Cases

File Name	Mean	Std. Dev.	Maximum	Minimum
Number of plate scattering points	17834	129	18083	17516
Number of volume scattering points	30969	136	31189	30583
Mean plate scattering angle	41.59	0.11	41.84	41.33
Mean volume scattering angle	41.04	0.08	41.24	40.92
Standard deviation plate scattering angle	3.88	0.01	3.90	3.84
Standard deviation volume scattering angle	3.07	0.00	3.07	3.06
Plate scattering signal	1.900	0.010	1.921	1.865
Volume scattering signal	10.756	0.078	10.877	10.530
Ratio volume scattering to plate scattering = RVP	5.660	0.037	5.769	5.576

Table 12 presents a statistical analysis of the calculations for the 87 cases selected by Teledyne. The mean values of the parameters will be used to normalize the results in many of the following analyses.

The number of scattering points in Table 12 can be used to assess the area of plate scattering and the volume of volume scattering. Note that these values will represent the outer edge of the tapered beams, not the half response width. The mean plate scattering area represents the transmitter beam foot print and is 0.03 X 0.03 X 17834 = 16.05 in^2. The mean volume scattering is 0.15 X 0.15 X 0.15 X 30969 = 104.5 in^3.

3.6.2 Data Extraction

The CMM data files were provided by Teledyne in three files containing some redundant data and some aborted measurements. The format is presented in Section A.2. The measurements were identified by two blocks of data:

1. Date, Program Name, Time, and Operator.
2. Serial Number, Part Number, Rev. Letter, Part Name, and four Comments.

Some measurements lacked the first block of data. There were also two slightly different measurement data formats, which required different processing. The following steps were taken to analyze the data:

1. Separate large files into numbered individual measurement files based on the first line number (N0025) for each measurement.
2. Extract the date and time parameters from files containing both data blocks (127 complete files lacked time and date).

3. Scan through each measurement file to look for the last line of the measurement (N0153 or N0154, depending on the format). 1497 complete data files were found, 1381 of which had date and time.

4. Extract the CMM measurements from each complete data file, using the format identified in Step 2. The CMM measurements saved included 9 production control parameters and 12 geometry parameters. The SIMCAL9 program was modified from SIMCAL8 to read the file name and these 21 parameters.

5. The redundant data files were eliminated by requiring unique date and time. The number of unique files was 676; 656 were complete.

6. 116 complete files without date and time remained. Requiring unique parameters reduced the number to 57 unique complete measurements lacking date and time. Thus, the total number of complete, unique data files was 713.

3.6.3 Results Based on Raw Serial Number

An examination of serial numbers and calibration dates gave some insight into the manufacturing process. The raw fork serial numbers were assigned as the forks were manufactured. The forks were then incorporated into fork assemblies, to which a new serial number was assigned, presumably in the order assembled.

Normally each fork was measured at least twice: first individually and then after being incorporated into a fork assembly (with mounting bracket). The assembly serial numbers have no particular relationship to the raw serial numbers, although some were assigned sequentially (see Figure 20). Some raw serial numbers had measurements with different recorded assembly serial numbers; these cases may represent data entry errors. The last raw fork measurements (through S/N 650) were made on February 8, 1997. The last assembly measurements (through raw S/N 643) were made on April 22, 1997. The assembly measurements of the highest raw serial numbers was not included in the measurement data files.

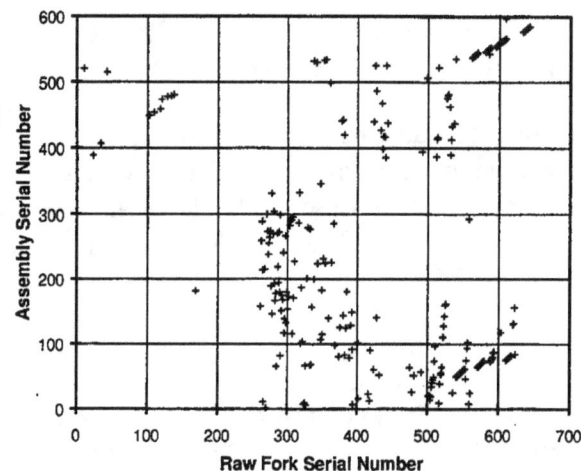

Figure 20. Relationship Between Assembly and Raw Serial Numbers

Of the 713 valid measurements, 585 had recorded raw fork serial numbers. These represented 297 unique forks. If the highest raw serial number of 650 represents the total number of forks produced, then measurements are available for slightly less than half the forks. Figure 21 shows the dates of the measurements, which more or less correspond to the raw serial number. The raw and assembly measurements were typically 1 or 2 months apart.

3.6.4 Calibration for All Measurements

For all subsequent analyses, four cases were removed from the 713 valid, unique measurements because the measurement data appeared to be inconsistent. Three cases (for forks with raw serial numbers 337, 424, and 634) were inconsistent with other measurements for the same fork. The fourth had no serial numbers. The number of cases is, therefore, at most, 709 in the following plots.

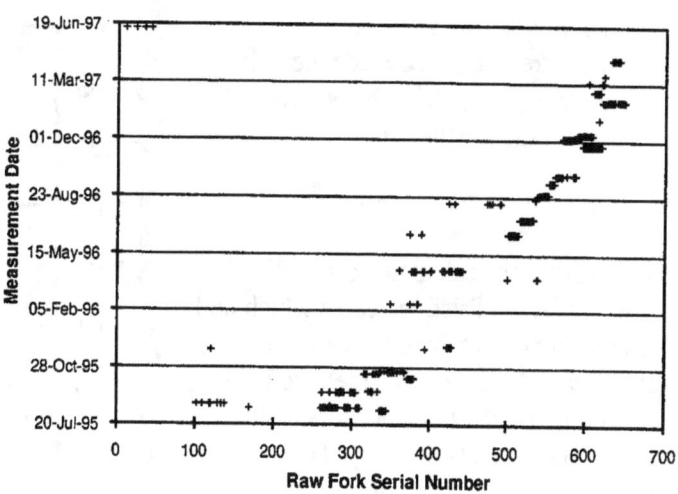

Figure 22. Measurement Date vs Raw Serial Number

Figure 22 shows the SIMCAL9 calculations of the volume to plate scattering ratio for each raw serial number. The following observations can be made:

1. All the measurements lie well within the ± 7% accuracy requirement. Most are within ± 3%.

2. Of the five early production forks, which were used in field tests, only one was outside the ± 3% range.

3. The end of the production run (S/Ns 620-650) shows a systematic drop in calculated

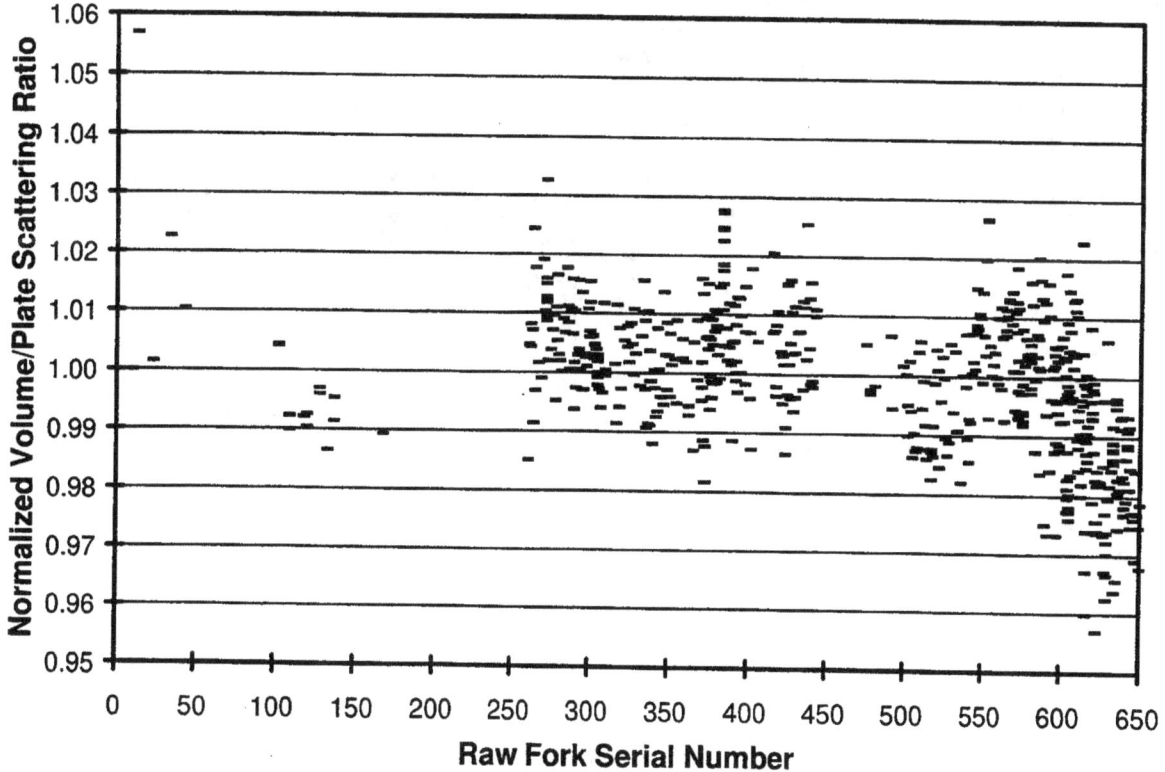

Figure 21. Results of SIMCAL9 Calculation vs. Raw Serial Number

calibration. Some of the measurements lie outside the ± 3% accuracy range that characterizes the rest of the production.

3.6.5 Calibration for 81 Forks

Figure 23 shows the calculated calibrations for the 81 of the 87 Teledyne cases which had raw serial numbers. Only one measurement is plotted for each fork; Teledyne selected the last measurement of the day for forks with multiple measurements on a day. The Teledyne cases were mostly from the middle of the production run.

The SIMCAL8 calibration distribution for the 87 Teledyne cases was plotted in Figure 19. Figure 24 shows the results for SIMCAL9. The spread of the distribution is reduced by more than a factor of two after the correction of the two errors in SIMCAL8. This change suggests that the fork geometry errors are small enough to permit the ± 7% accuracy to be readily met. Section 4.1 will address the possible contributions of head errors to the unit-to-unit calibration error.

3.6.6 CMM Calibration Consistency

Figure 25 shows how the calculated calibrations varied for most of the forks with more than one measurement. The variations depend upon the accuracy of the determination of the scattering geometry and also the statistical variations of the calculation method.

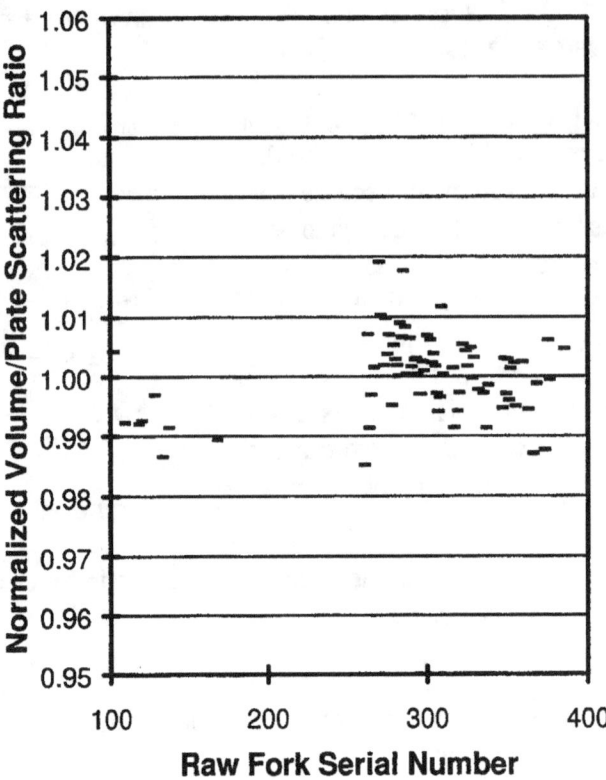

Figure 23. Figure 22 for 81 Teledyne Forks

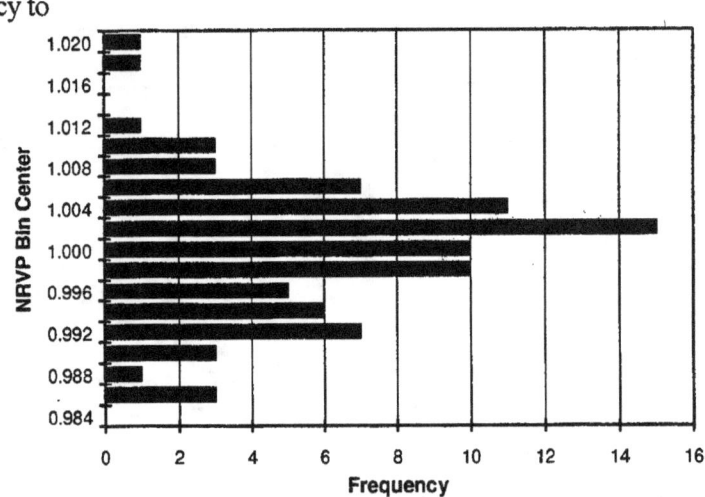

Figure 24. Distribution of Normalized Ratio of Volume to Plate Scattering

The spread in measurement varies from one fork to the next, but is usually less than 2% for forks with many measurements. The maximum spread is about 5%, which occurs for only a few forks that have only one point defining the large spread. These points might simply be measurement outliers. Apart from such outliers, the variance in calculated calibration has little impact on meeting the ± 7% accuracy criterion. The analysis of Table 10 suggests that the observed variance is related to measurement errors rather than simulation uncertainties.

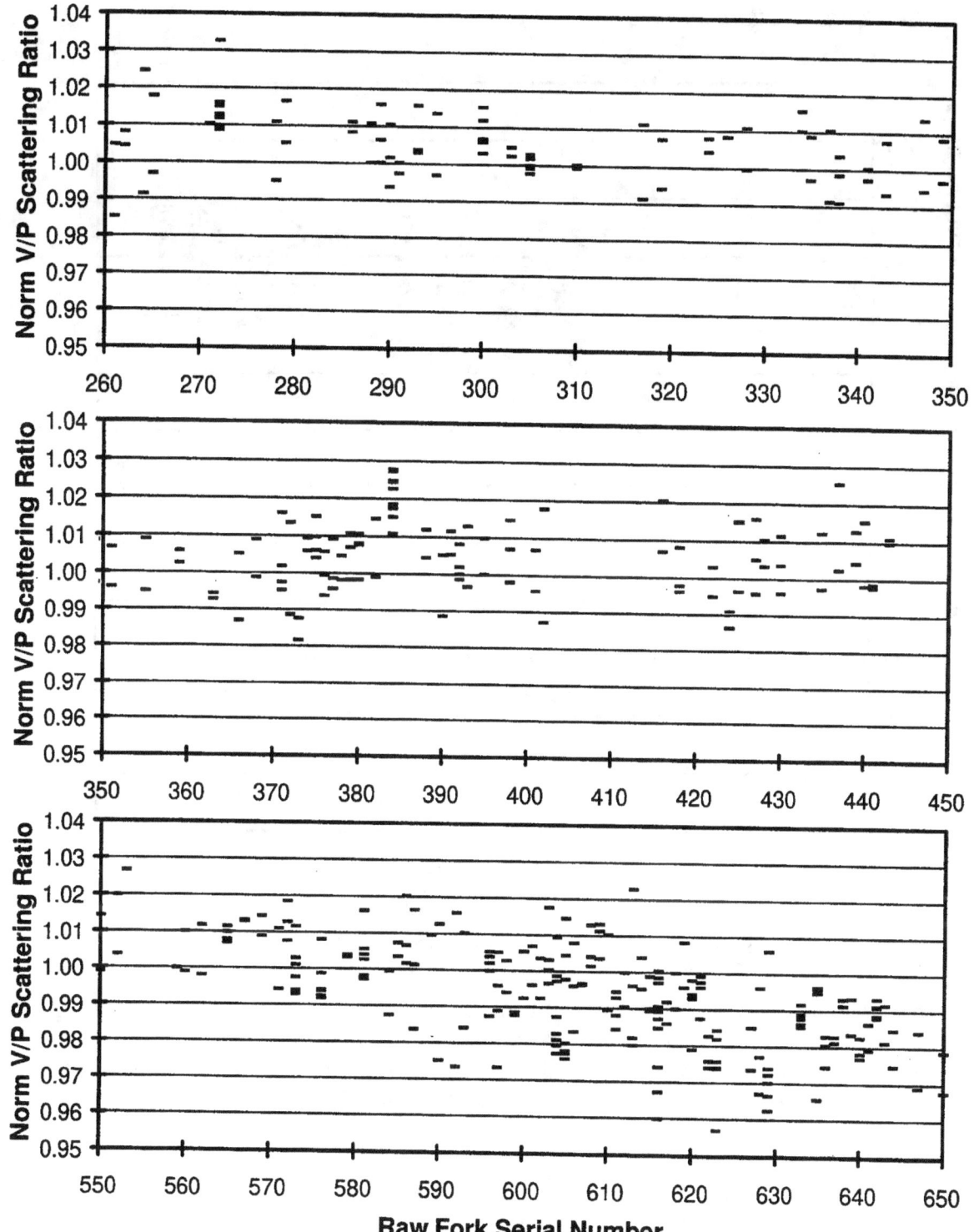

Figure 25. Calibrations for Duplicate Measurements

3.7 PRODUCTION CONSISTENCY

Existing CMM measurements can be used to examine the consistency of fork geometry through much of the production process; however, some gaps do exist.

3.7.1 Compound Scattering Angle

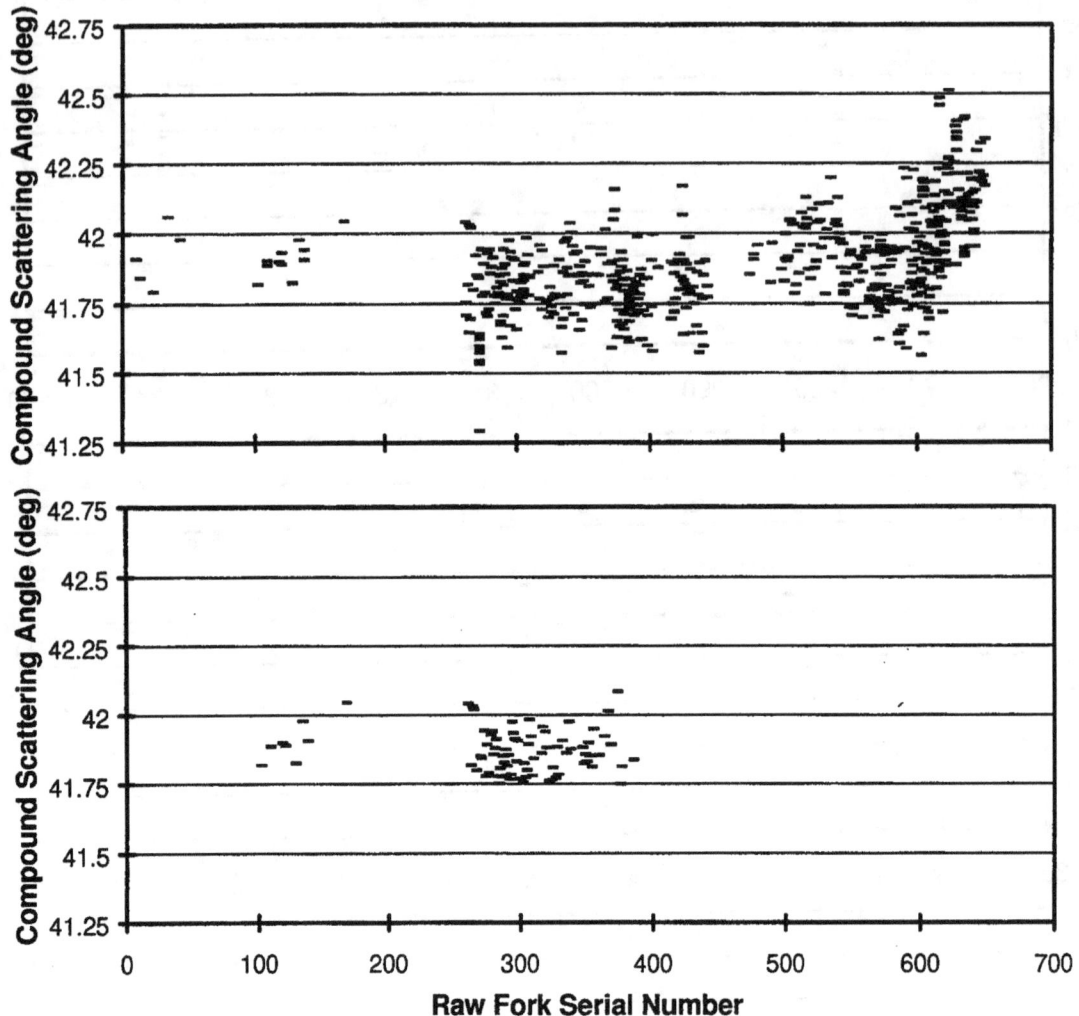

Figure 26. Compound Scattering Angle vs Raw Serial Number
Top = All Cases; Bottom = 81 Teledyne Cases

Figure 26 shows the measurements of the fork compound scattering angle through the production run. Many measurements (top) lie outside the production limits of 42.00 ± 0.25 degrees. In general, the compound angles are biased below 42 degrees. However, the bias appears to increase for the last 30 forks, the same ones that gave lower calculated calibrations in Figure 22.

All the 81 measurements selected by Teledyne (bottom of Figure 26) lie within the production limits. The goal of passing the accuracy test explains why the selected Teledyne measurement was always last in a sequence. The measurement was repeated until the results met the compound angle specification. The measurement fluctuations were enough to pass forks with a typical compound angle measurement that might be out of specification.

3.7.2 In-Plane Scattering Angle

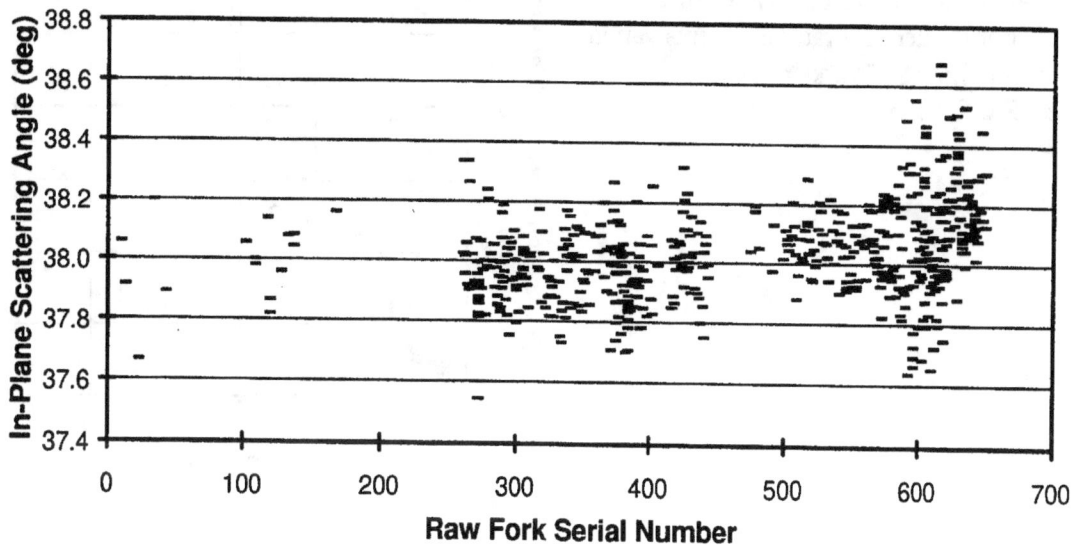

Figure 27. In-Plane Scattering Angle vs Raw Serial Number

Figure 27 shows how the in-plane scattering angle varied through the production run. Again the last 30 forks showed an increased scattering angle. The tolerances on in-plane angle are 38.0 ± 0.5 degrees and only a few measurements are outside the limits.

3.7.3 Calibrator Intercepts

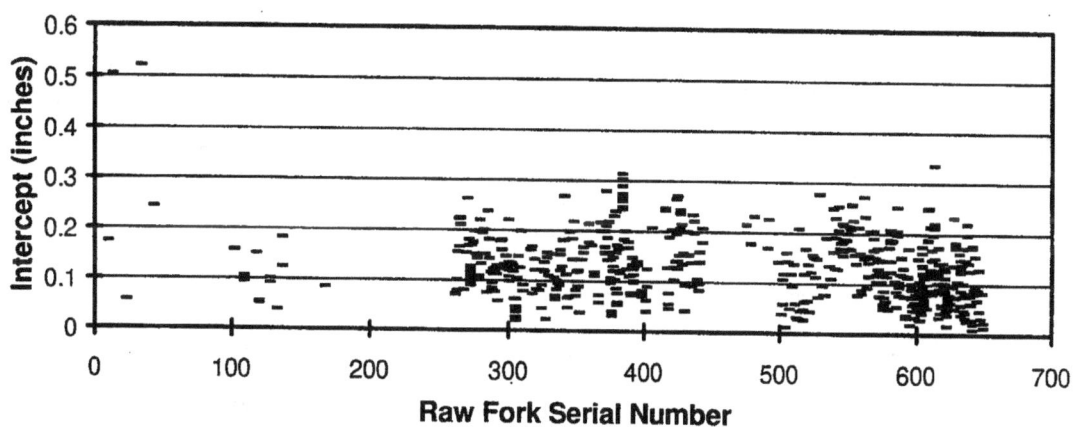

Figure 28. Intercept Spacing vs Raw Serial Number

Figure 28 shows how the distance between the transmitter beam and receiver beam intercepts with the calibration plate varied with the production run. Virtually all the forks met the 0.30 inch limit. The most notable exceptions were the two early production units with intercept spacing > 0.50 inches.

Figure 28 plots the data from the part of the CMM program used to calculate compliance with the specification. The intercept information also was output at the end of the program where the complete fork geometry was provided. The intercept spacing based on the latter values differed somewhat from the data shown in Figure 28. Figure 29 shows a scatter plot comparing the two results. Differences as large as 0.05 inches are noted. Perhaps the differences are due to round-off errors.

The intercept details are presented in Figures 30 through 32. Figures 30 and 31 show how the receiver and transmitter calibrator intercepts varied. Figure 32 shows how the Z and Y intercept differences (Rx-Tx) varied.

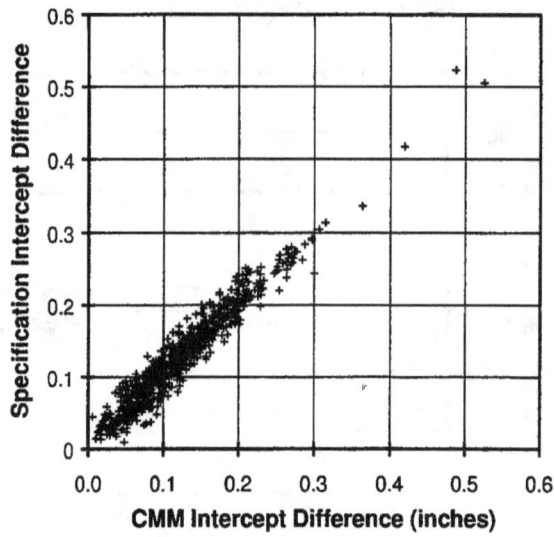

Figure 29. Comparison of Two Intercept Difference Values

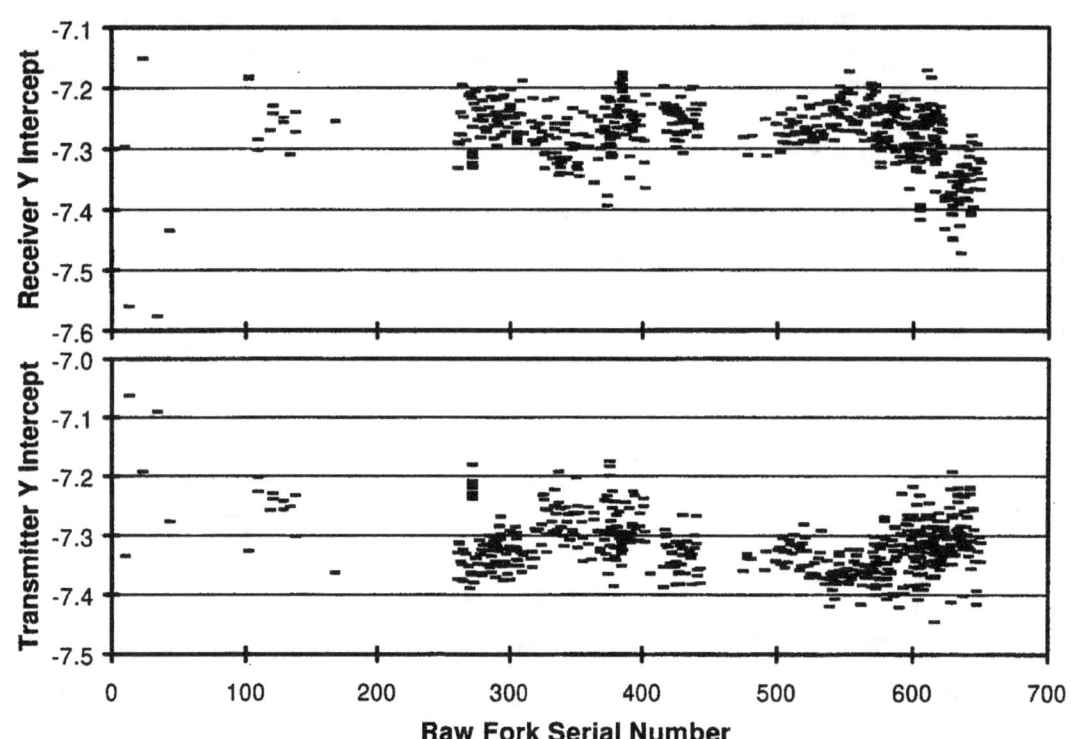

Figure 30. Receiver & Transmitter Calibrator Y Intercept vs Raw Serial Number

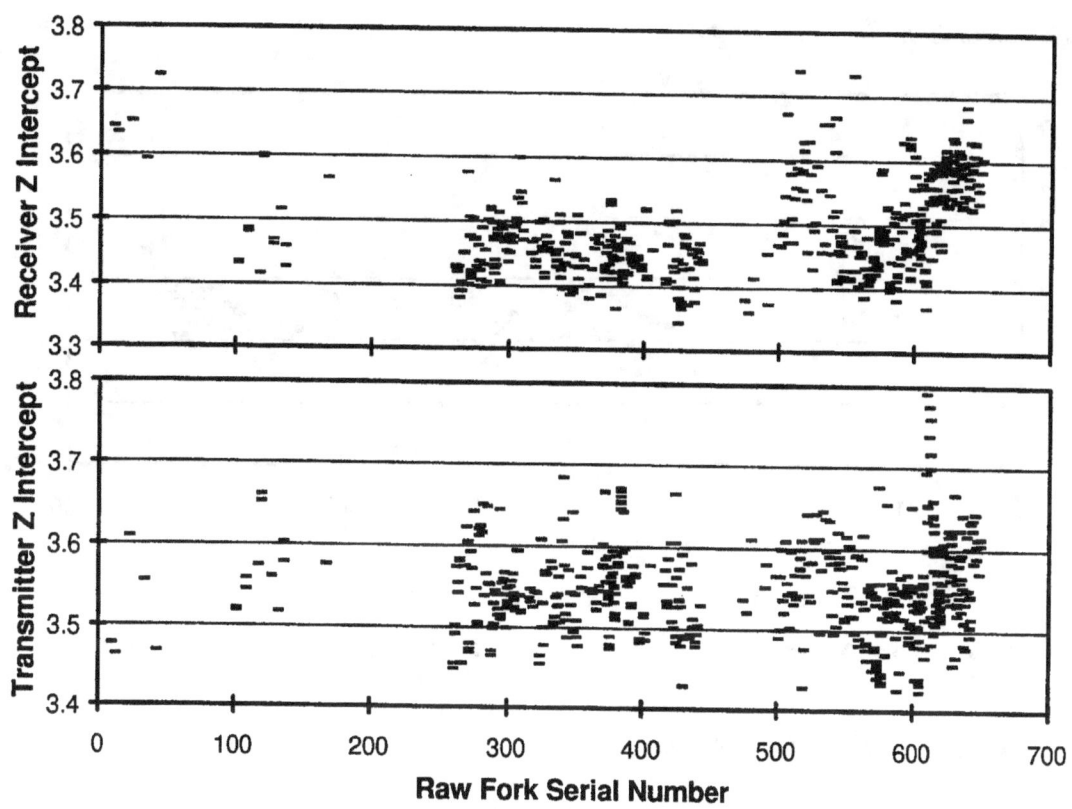

Figure 31. Receiver & Transmitter Calibrator Y Intercept vs Raw Serial Number

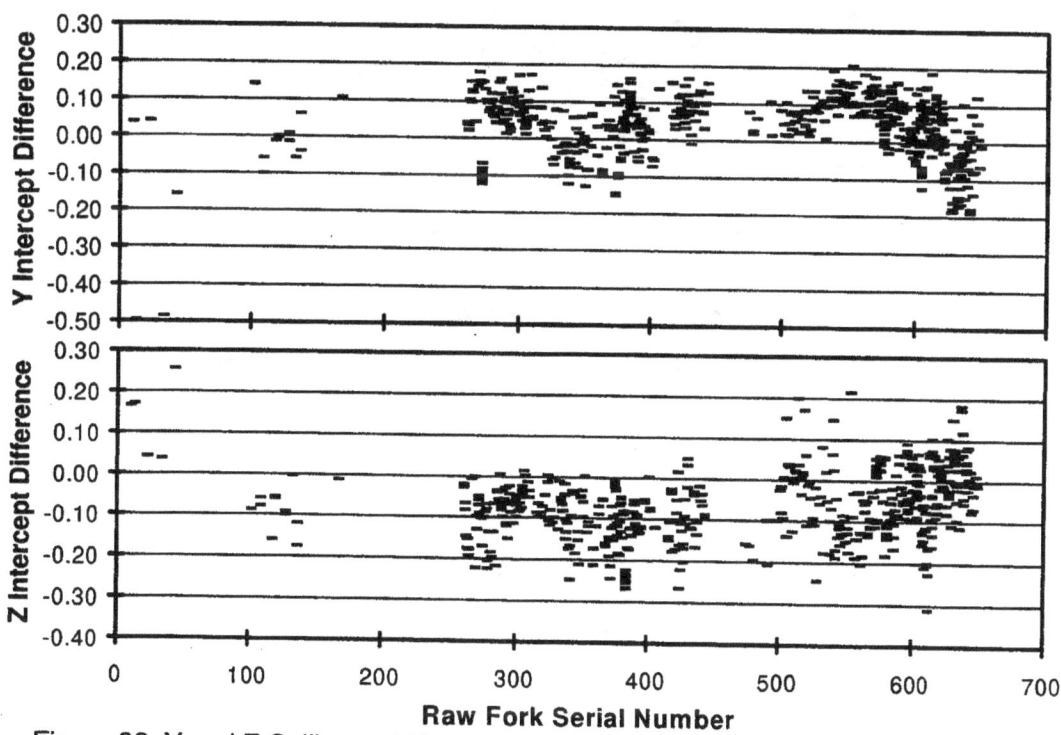

Figure 32. Y and Z Calibrator Y Intercept Rx-Tx Differences vs Raw Serial Number

3.7.4 Receiver Out-Of-Plane Angle

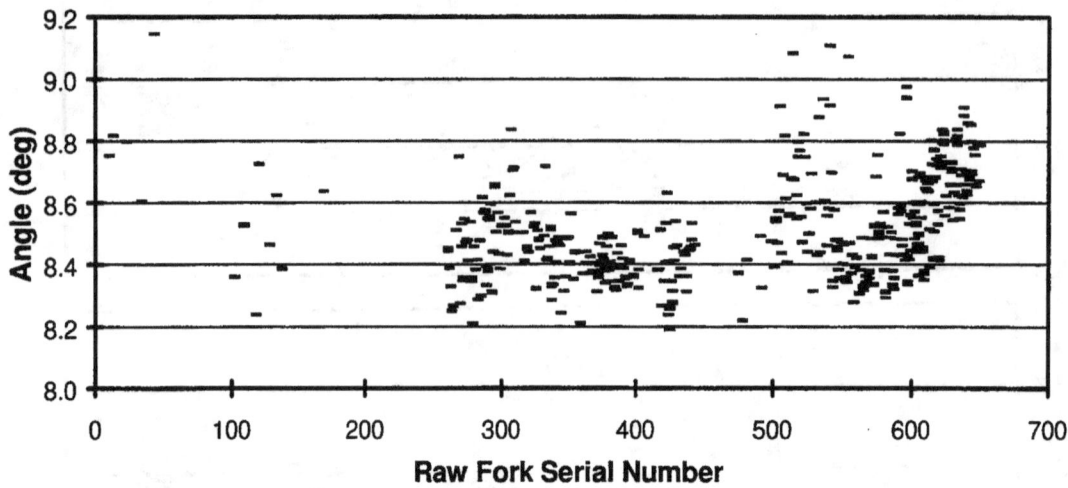

Figure 33. Receiver Out-Of-Plane Angle vs Raw Serial Number

Figure 33 shows how the receiver out of-plane angle varied through the production run. The tolerance limit was 8.7 ± 1.0 degrees, which was easily met.

3.7.5 Transmitter Out-Of-Plane Angle

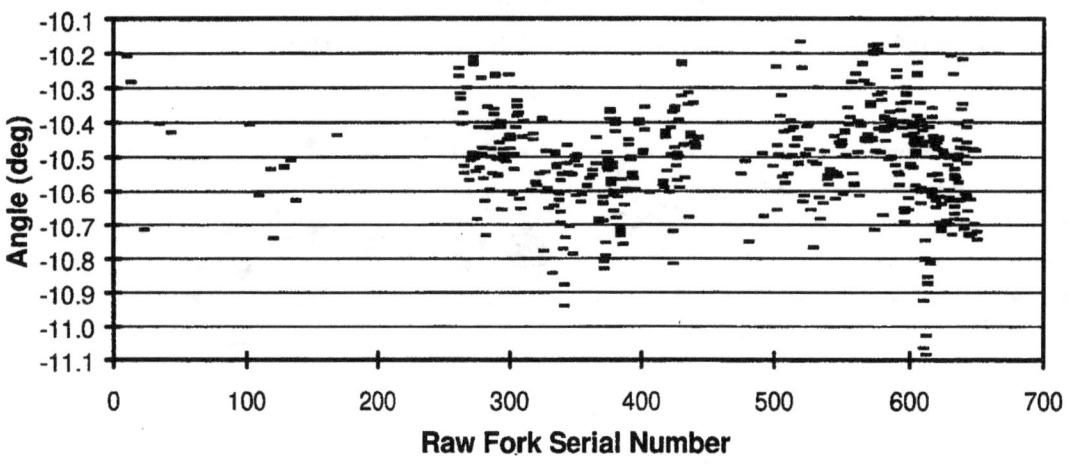

Figure 34. Transmitter Out-Of-Plane Angle vs Raw Serial Number

Figure 34 shows how the receiver out of-plane angle varied through the production run. The tolerance limit was -10.5 ± 1.0 degrees, which was easily met.

3.7.6 Head Mount Angles

Figure 35 shows how the three head mount angles varied through the production run. The Θ angles (top for receiver, middle for transmitter) show the tilt in the fork plane of the normal to the two head mounts, which is nominally 90 ± 19 degrees, i.e., each beam tilted down by 19 degrees in the plane of the fork. Most of the head mounts are within ± 1.5 degrees of the nominal value. Note that the difference in these angles is the in-plane angle of Section 3.7.2. The y-z coordinate system orientation was defined to have the normal to the transmitter mount to be in the x-y plane (i.e., the angle out of the x-y plane, $\Phi_t = 0.00$). The receiver normal out-of-xy-plane angle, Φ_r (bottom of Figure 32) then indicates the alignment consistency of the two heads and the calibrator. The same tolerance is observed as for the other two head mount angles, namely $\Phi_r = 0.0 \pm 1.5$ degrees.

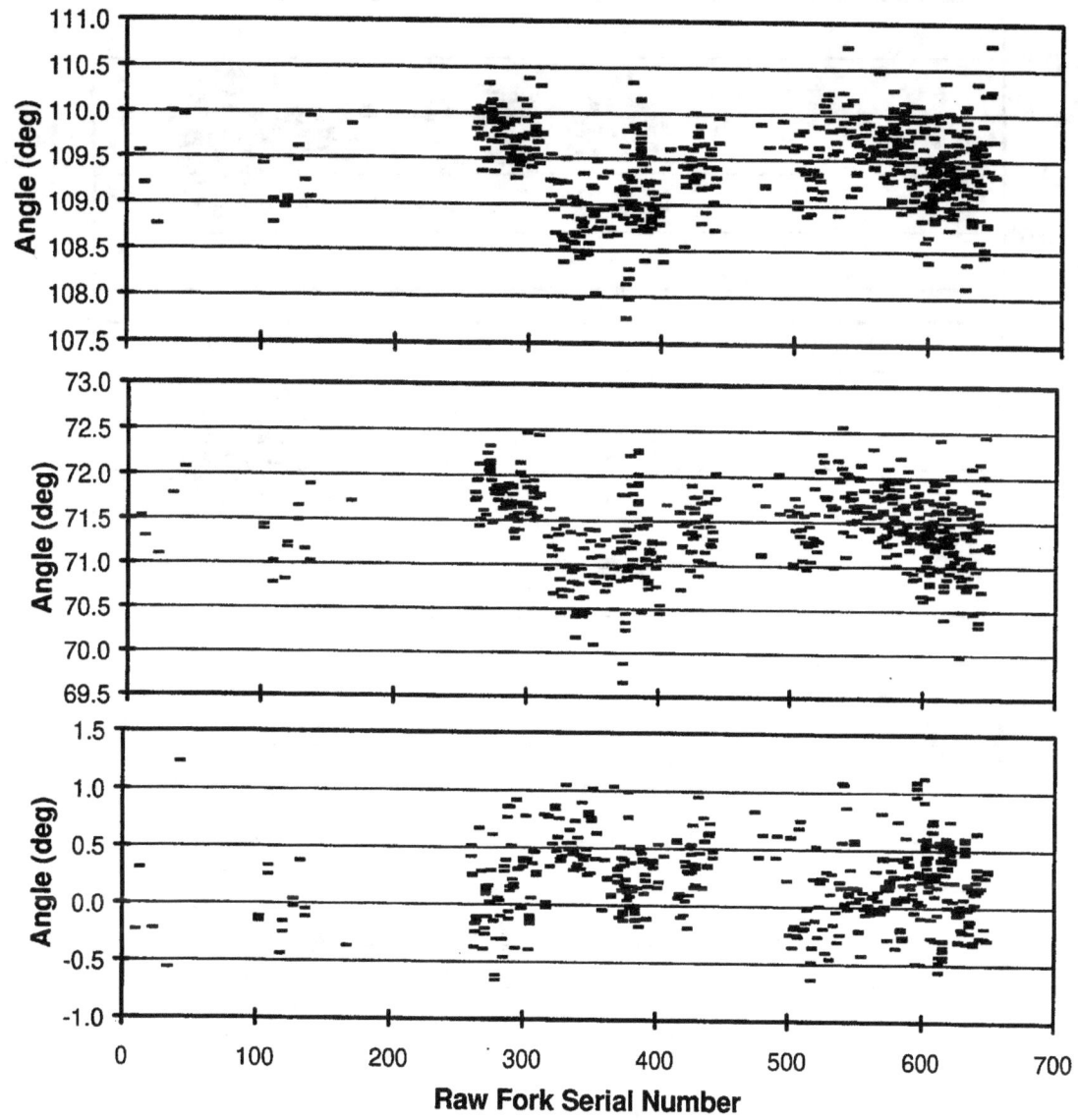

Figure 35. Fork Mount Angles vs Raw S/N: Top = Θ_r; Middle = Θ_t; Bottom = Φ_r

3.7.7 Distances

Figure 36 shows how three fork distances varied through the production run. The transmitter to receiver spacing (top) was usually well controlled by the welding jig to less than ±0.1 inches. The early bent forks showed greater variation, as would be expected. Most of the forks with raw serial numbers 100 to 140 had very consistent, low values. The calibrator-transmitter (middle) and calibrator-receiver (bottom) spacings showed greater variability, presumably because of slight tilts in the calibrator mount. Becasue their sum is roughly constant, the calibrator-transmitter and calibrator-receiver distances variance are anticorrelated.

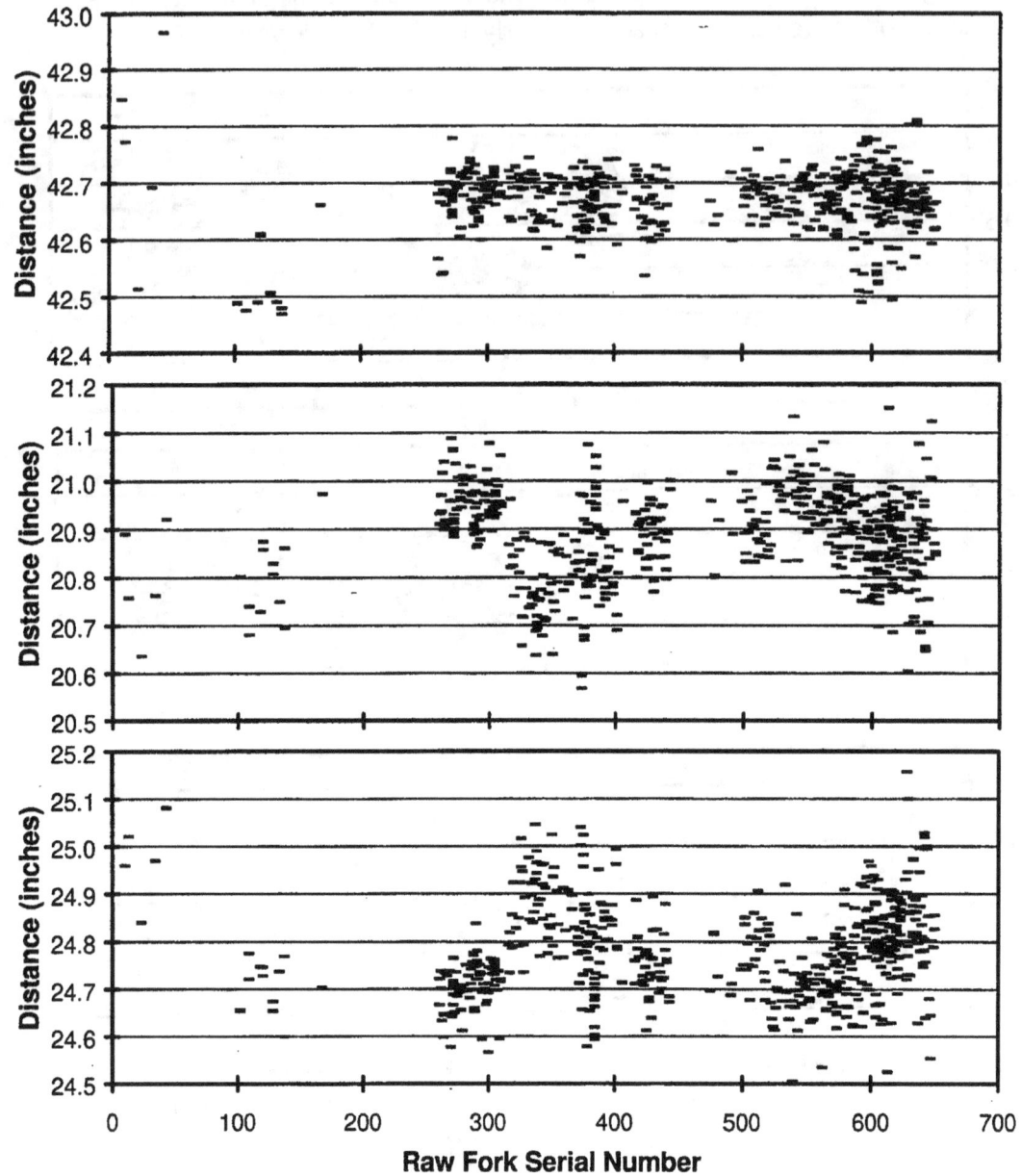

Figure 36. Fork Distances vs Raw S/N: Top = Receiver to Transmitter, Middle = Transmitter to Transmitter Calibrator Intercept; Bottom = Receiver to Receiver Calibrator Intercept

The positions of the receiver and transmitter mounts are shown in Figures 37 through 39. Considerable correlation is shown in these values. Because slight tilts of the calibrator can move the x-axis origin, the mount X locations vary together. Because the y-axis and z-axis origins are set by the line connecting transmitter and receiver mount, their Y and Z coordinates are anticorrelated.

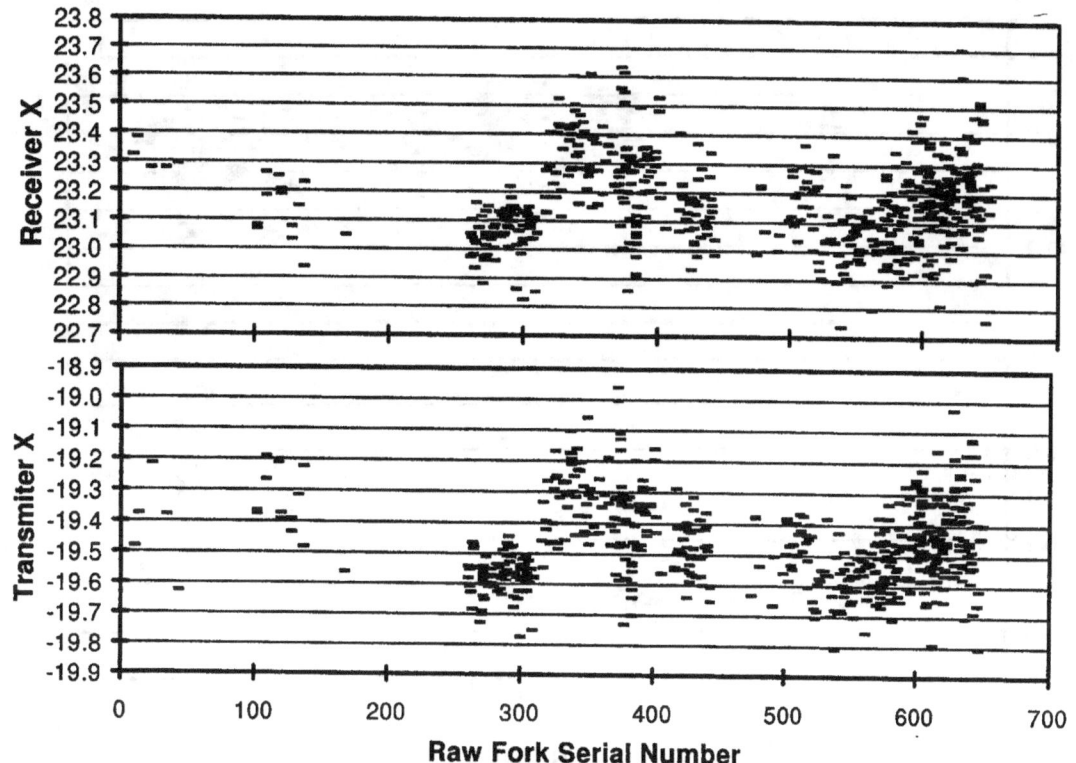
Figure 37. Receiver & Transmitter X vs Raw Serial Number

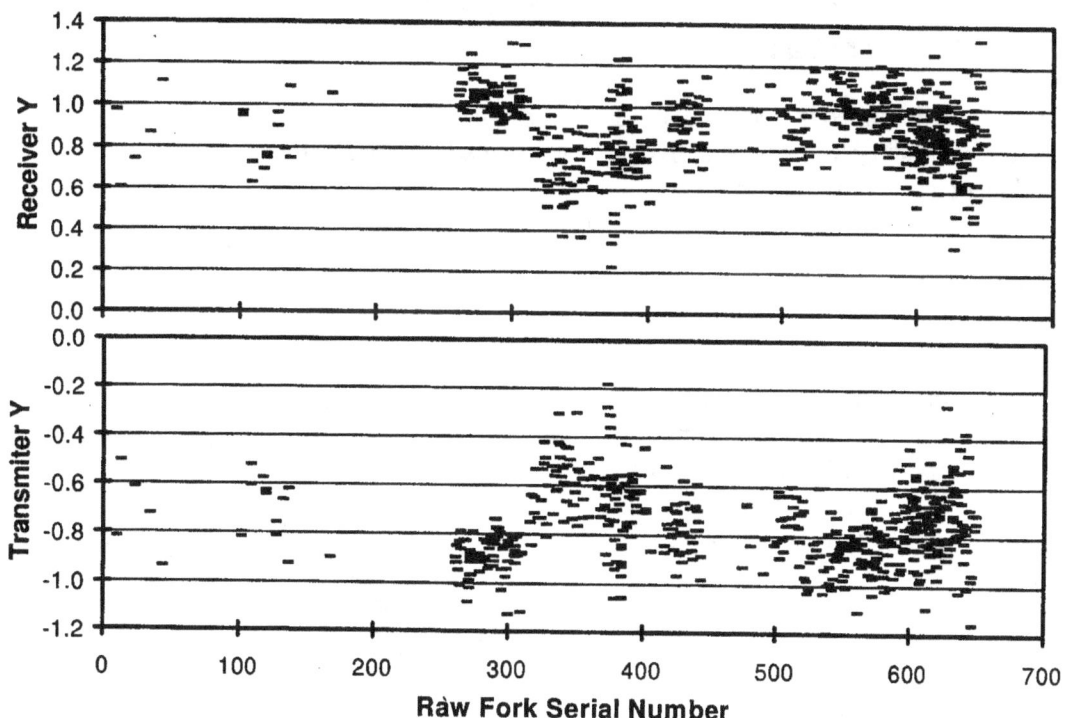
Figure 38. Receiver & Transmitter Y vs Raw Serial Number

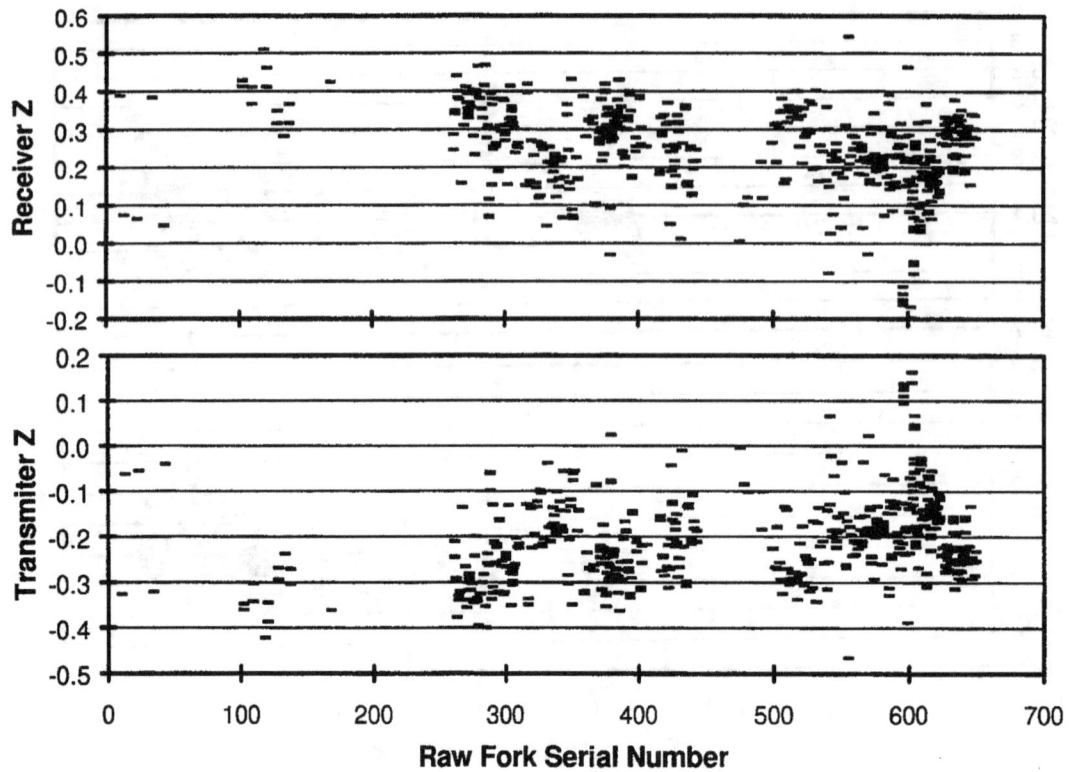

Figure 39. Receiver & Transmitter Z vs Raw Serial Number

3.7.8 Parameters after Head Corrections

The head geometry described in Section 3.3 is used to calculate the actual beam origin locations and leads to many changes in geometry beyond the fork parameters described above. Some of these changes will be presented in this section.

Perhaps the most significant change is the modification of the distance between the transmitter and receiver calibrator intercepts. Figure 40 compares the intercepts before and after the head corrections. A variation of ± 0.05 inches is observed. The significance of these changes is that calculating the beam intercept distance based on fork geometry alone does not correctly specify the scattering geometry. However, since the variance of ± 0.05 inches is much less than the production limit of ± 0.30 inches, the errors introduced are relatively unimportant.

Figure 41 shows the detailed intercept changes resulting from the head corrections. The Y coordinate changes are of both signs and typically smaller than the Z changes. The Z changes are

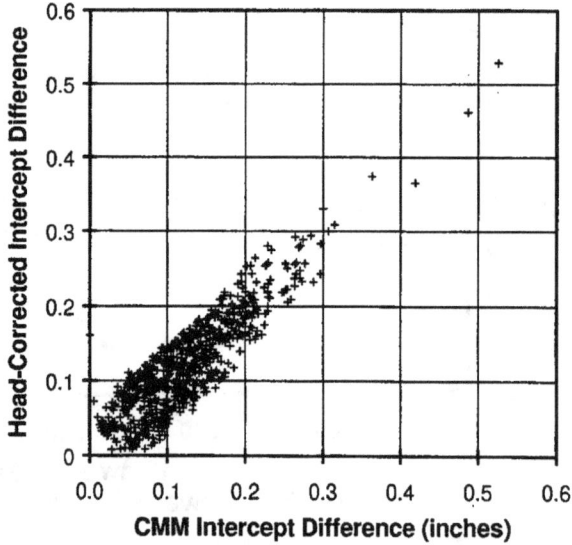

Figure 40. Comparison of Intercept Differences Before/After Head Corrections

positive and result from the difference in out-of-plane angle of the transmitter and receiver. Raising the beam origin above the head mounts effectively moves the beams closer to the calibrator and results in a systematic change in Z coordinate displacement.

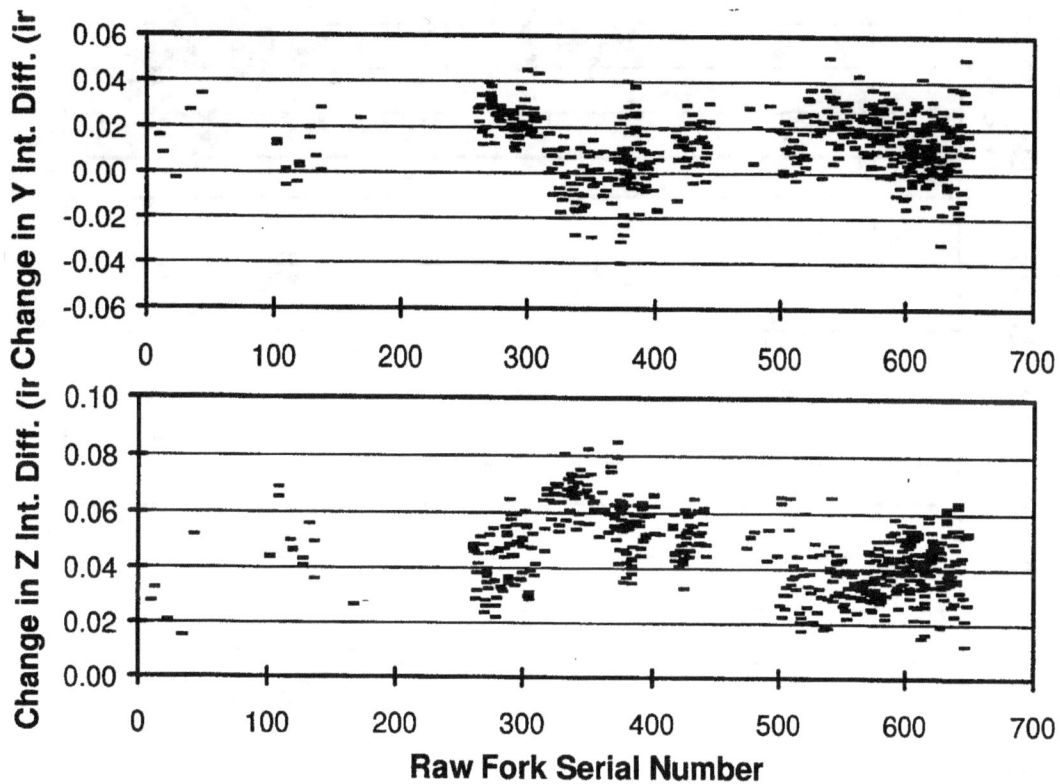

Figure 41. Changes In Rx-Tx Intercept Differences from Head Corrections

The head corrections lead to increased distances between beam origins and calibrator and beam origins, compared to the nominal fork values plotted in Figure 36. Figure 42 shows the distances resulting from using beam origins rather than head mount locations.

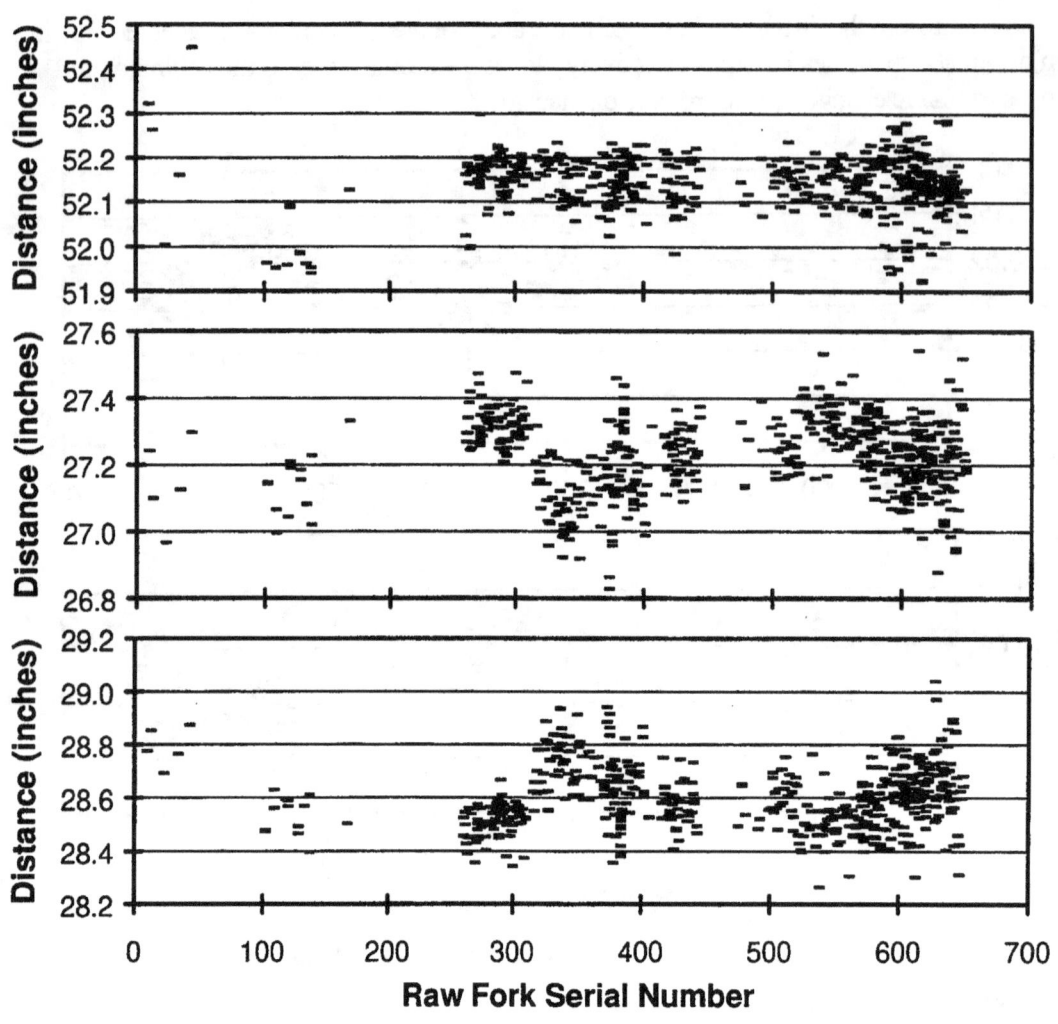

Figure 42. Beam Distances vs Raw S/N: Top = Receiver to Transmitter; Middle = Transmitter to Transmitter Calibrator Intercept; Bottom = Receiver to Receiver Calibrator Intercept

3.8 CALCULATED CALIBRATION VERSUS MEASURED PARAMETERS

The influence of the variances in the fork measurements on the calibration can be assessed by plotting the parameters of the last section against the normalized calculated volume/plate scattering ratio. These plots are presented in this section in Figures 43 through 50.

3.8.1 Ratio Volume to Plate Scattering

The scattermeter calibration variation depends upon the variation in the ratio of volume to plate scattering, which is examined in the following scatter plots.

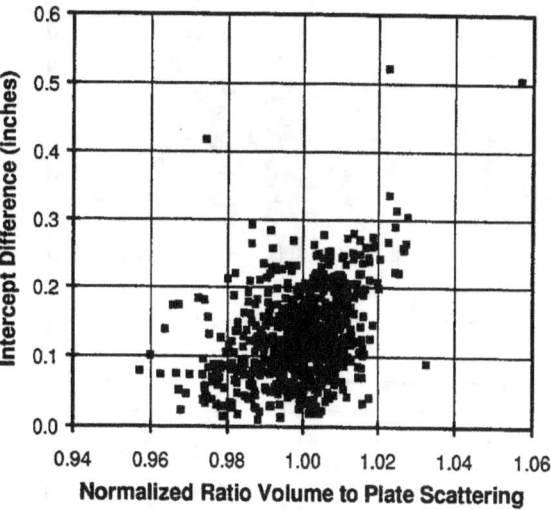

Figure 44. Rx-Tx Calibrator Intercept Difference vs NRVP

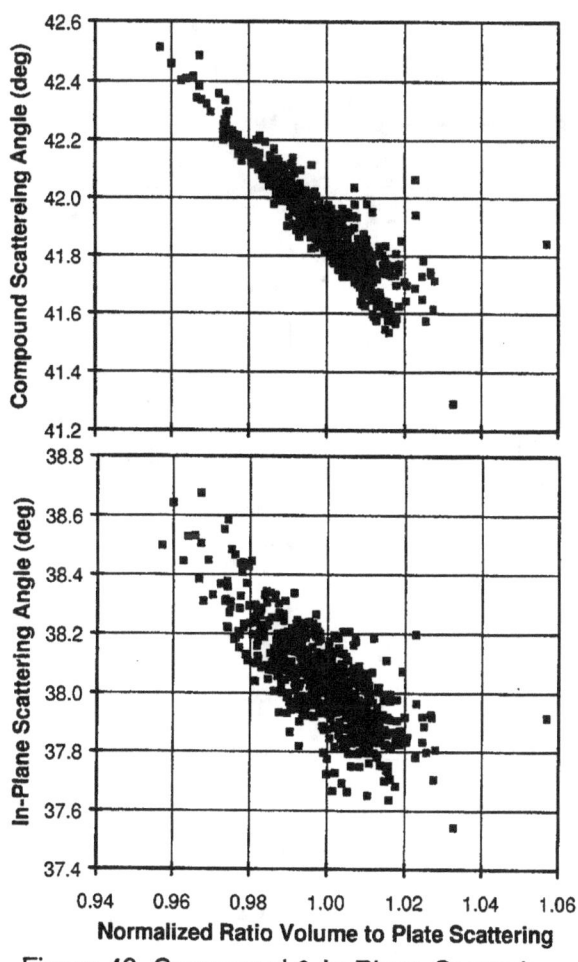

Figure 43. Compound & In-Plane Scattering Angle vs NRVP

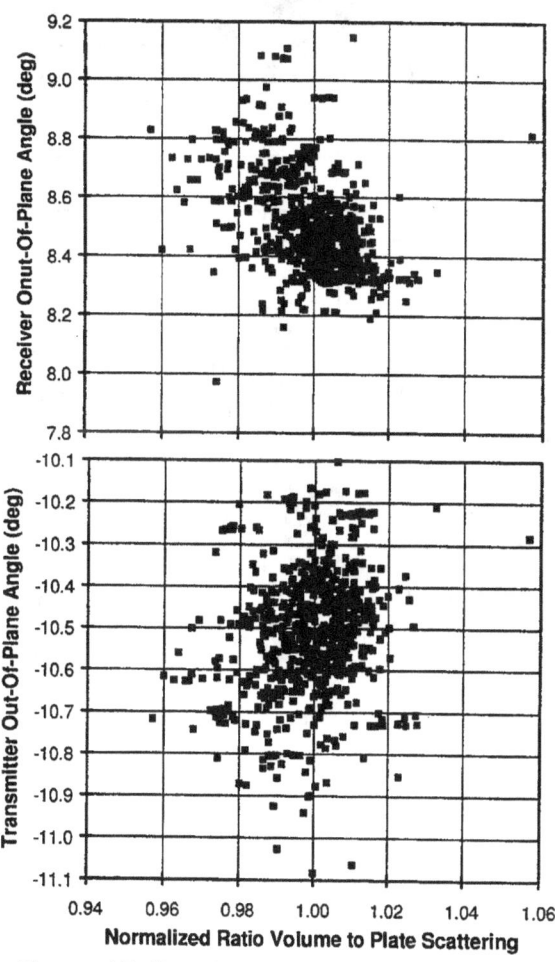

Figure 45. Receiver & Transmitter Out-Of-Plane Angles vs NRVP

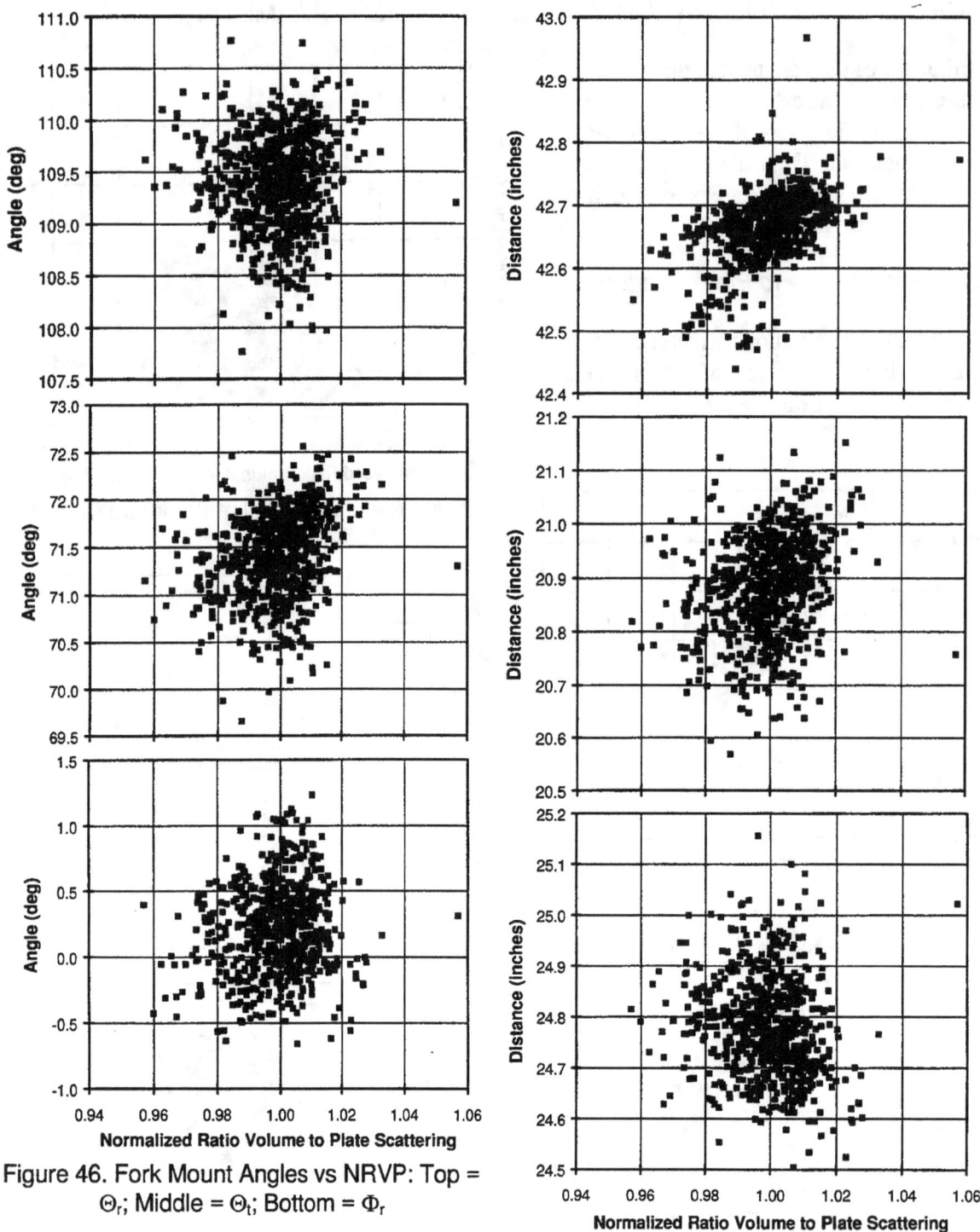

Figure 46. Fork Mount Angles vs NRVP: Top = Θ_r; Middle = Θ_t; Bottom = Φ_r

Figure 47. Distances vs NRVP: Top = Rx-Tx; Middle = Tx-Cal Int.; Bottom = Rx-Cal Int.

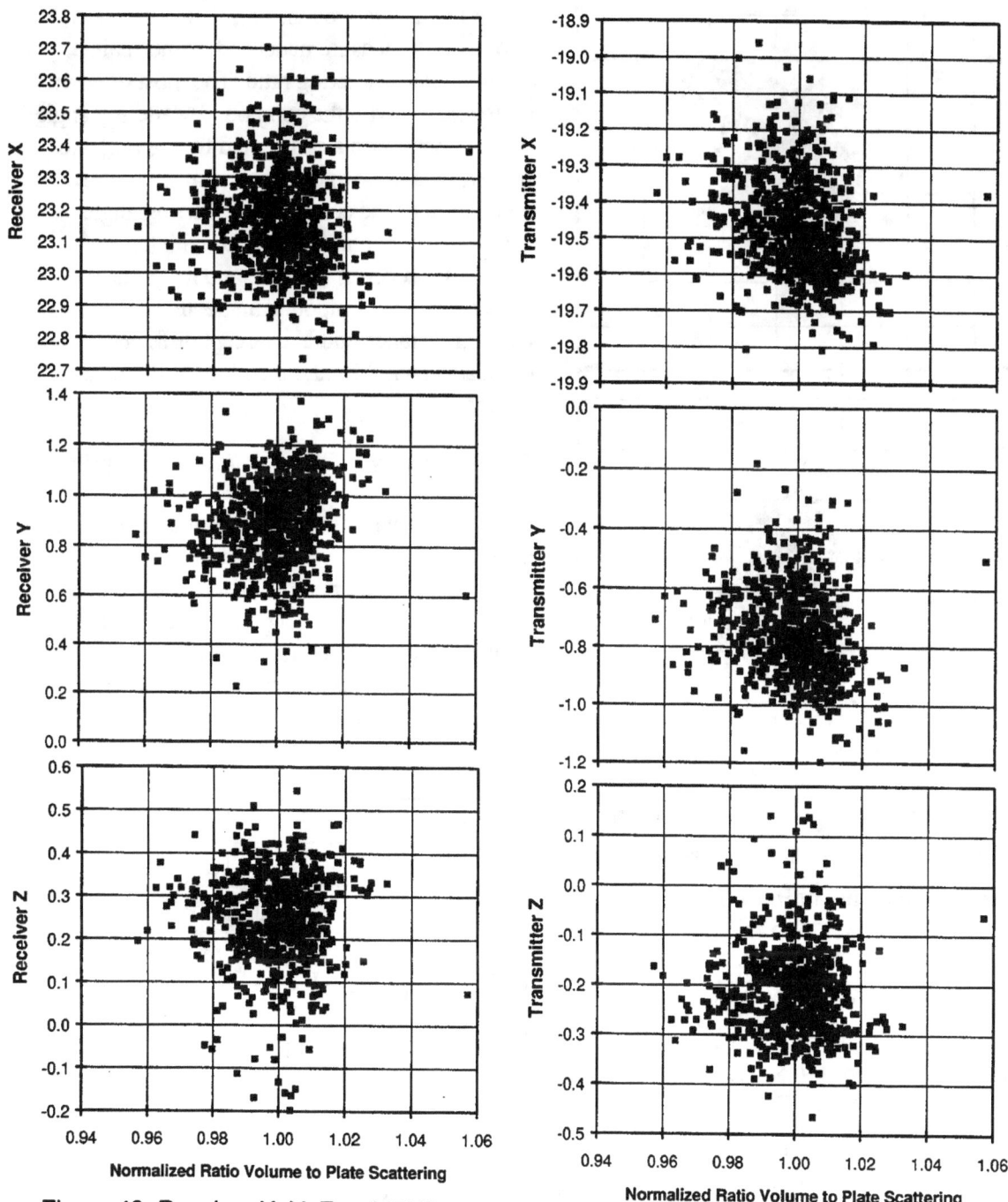

Figure 48. Receiver X, Y, Z vs NRVP

Figure 49. Transmitter X, Y, Z vs NRVP

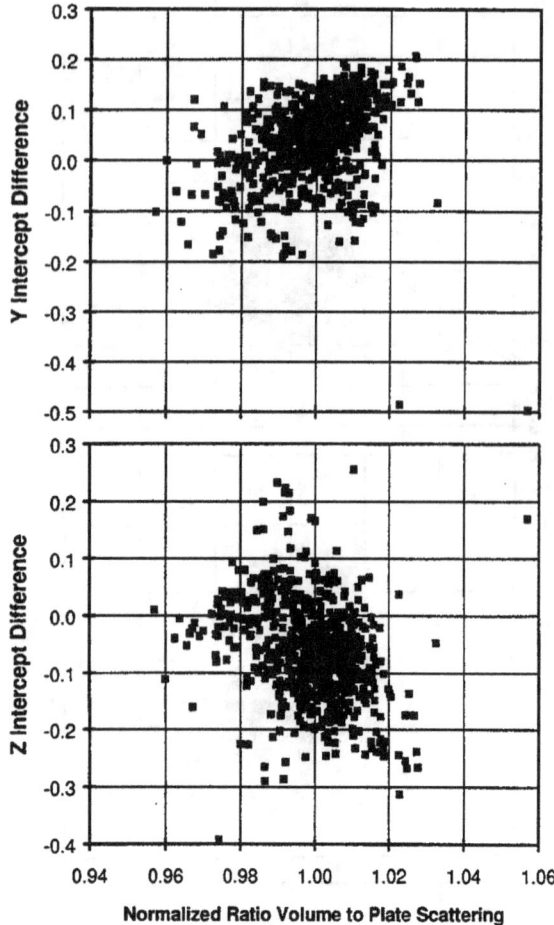

Figure 50. Intercept Differences vs NRVP

Table 13. Summary of Parameter Correlation with NRVP

Parameter	Correlation
Compound Sct. Ang.	Very Strong, Negative
In-Plane Sct. Ang.	Strong, Negative
Intercept	Weak, Positive
Rx Out-Plane Angle	Weak, Negative
Tx Out-Plane Angle	None
Rx Mount Θ	None
Tx Mount Θ	Weak, Positive
Rx Mount Φ	None
Rx-Tx Spacing	Moderate, Positive
Rx-Cal Spacing	None
Tx-Cal Spacing	None
Rx/Tx X	None/Weak, Negative
Rx/Tx Y	Weak, Positive/Negative
Rx/Tx Z	None
Y Intercept Diff.	Moderate, Positive
Z Intercept Diff.	Weak, Negative

Table 13 summarized the correlation between the various fork measurements and the normalized volume/plate scattering ratio. The most dramatic correlation is with the compound scattering angle. Strong correlation also is noted for the projected scattering angle, which captures much of the correlation of the compound scattering angle. Moderate correlation was noted with the transmitter-receiver distance and the Y intercept difference. The compound angle, transmitter-receiver distance and Y intercept difference correlation will be examined further in the next section.

3.8.2 Separate Plate/Volume Scattering

Becasue the plate and volume scattering depend differently on the measured parameters, it might be easier to understand the calibration variations if the two types of scattering are studied separately. In this section, the plate and volume scattering parameters have been normalized to the mean values (see Table 12) for the 87 cases selected by Teledyne.

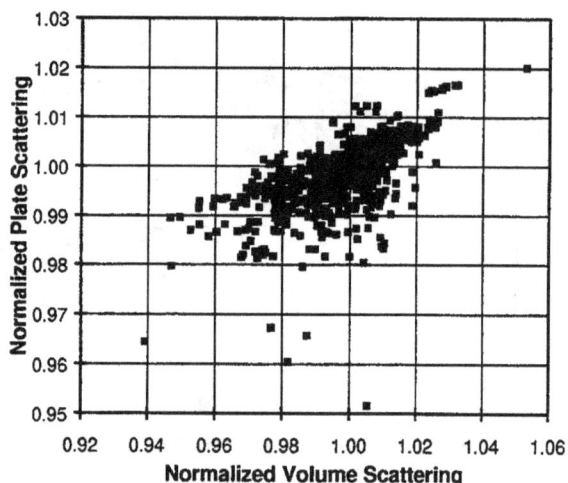

Figure 51. Correlation Between Plate and Volume Scattering

Figure 51 shows strong correlation between the calculated plate and volume scattering. Apart from outliers, plate scattering shows less fractional variance than volume scattering.

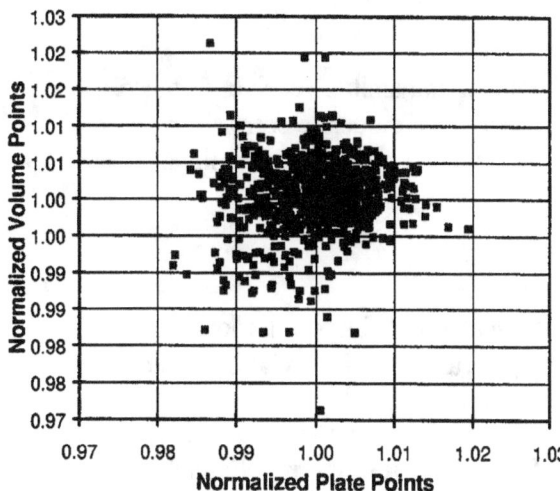

Figure 52. Correlation Between Number of Plate and Volume Scattering Points

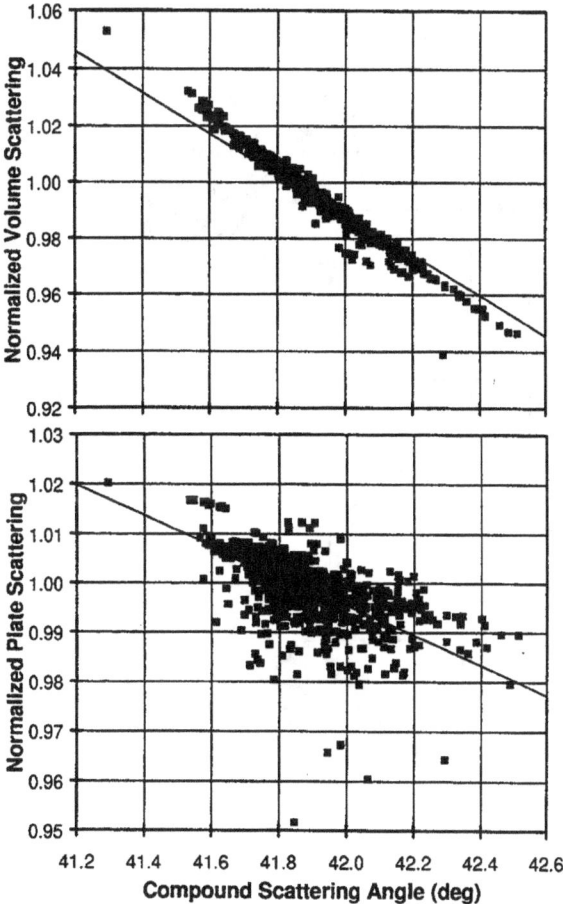

Figure 53. Correlation Between Volume/Plate Scattering and Compound Scattering Angle

Figure 52 shows that the number of volume scattering points varies similarly to the number plate scattering points. The number of scattering points corresponds roughly to the overlap between transmitter and receiver beams.

3.8.2.1 Compound Angle

Figure 53 shows the correlation between volume and plate scattering with compound scattering angle. These plots also show the modeled fractional dependence of the scattering on scattering angle (-0.07 per degree for volume and -0.03 per degree for plate). The calculated variance in scattering with compound scattering angle is close to these assumed values. The volume scattering turns out to have a slightly greater dependence. The strong variation in volume scattering with scattering angle is clearly the dominant factor in defining the calculated calibration variance.

3.8.2.2 Transmitter-Receiver Distance

Figure 54 shows the dependence of volume and plate scattering on transmitter-receiver distance. Only the volume scattering shows significant correlation. Consequently, the moderate correlation observed for normalized ratio of volume to plate scattering (NRVP) in Figure 47 must be predominantly caused by volume scattering, not plate scattering.

3.8.2.3 Intercept Differences

Since intercept differences will form the basis of much of the analysis of Chapter 4, more intercept difference data will be presented here than strictly justified by the amount of correlation observed.

Figures 55 and 56 show the effect of Y and Z intercept differences on volume and plate scattering, respectively. Relatively little correlation is noted in Figure 55 for volume scattering. Figure 56 for plate scattering shows clearer correlation, with opposite signs for the Y and Z components.

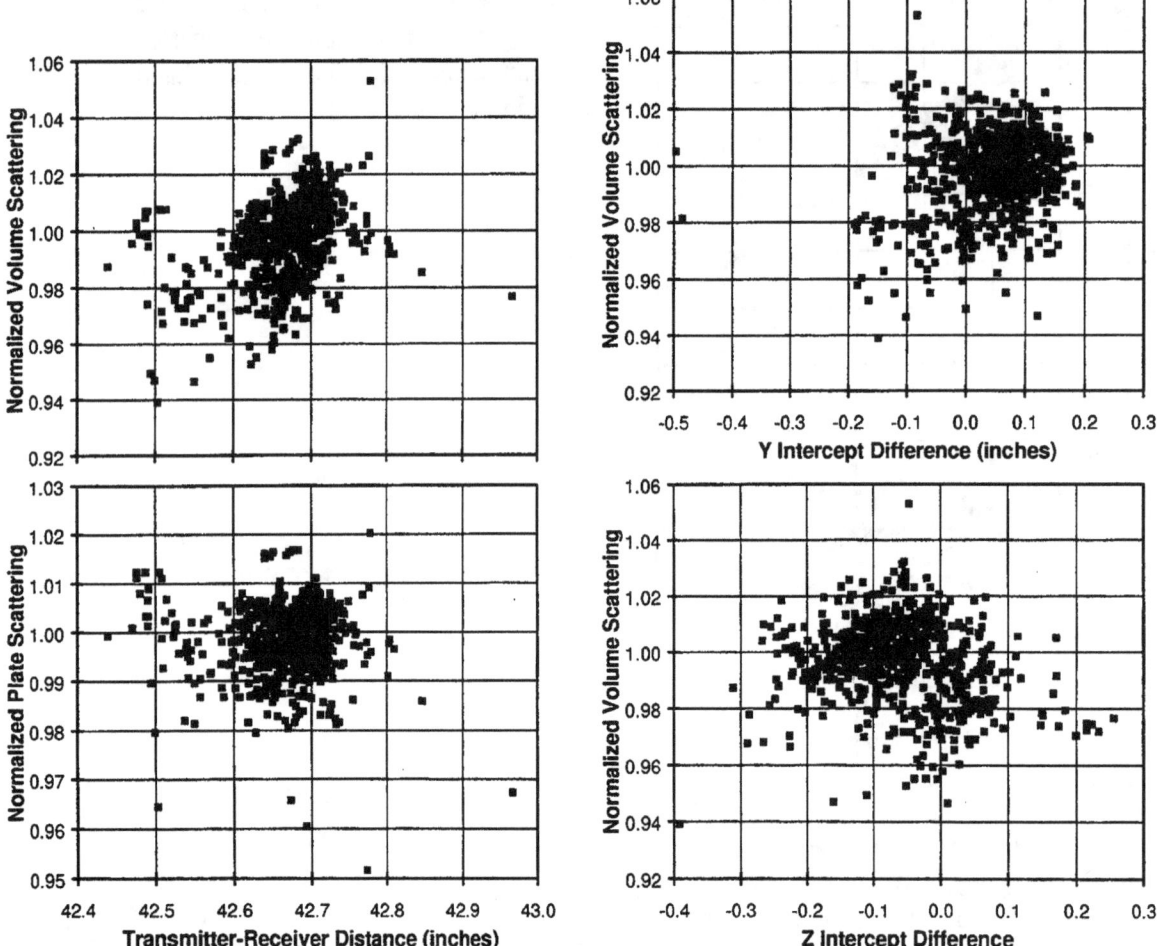

Figure 54. Correlation Between Volume/Plate Scattering and Transmitter-Receiver Distance

Figure 55. Dependence of Volume Scattering on Intercept Differences

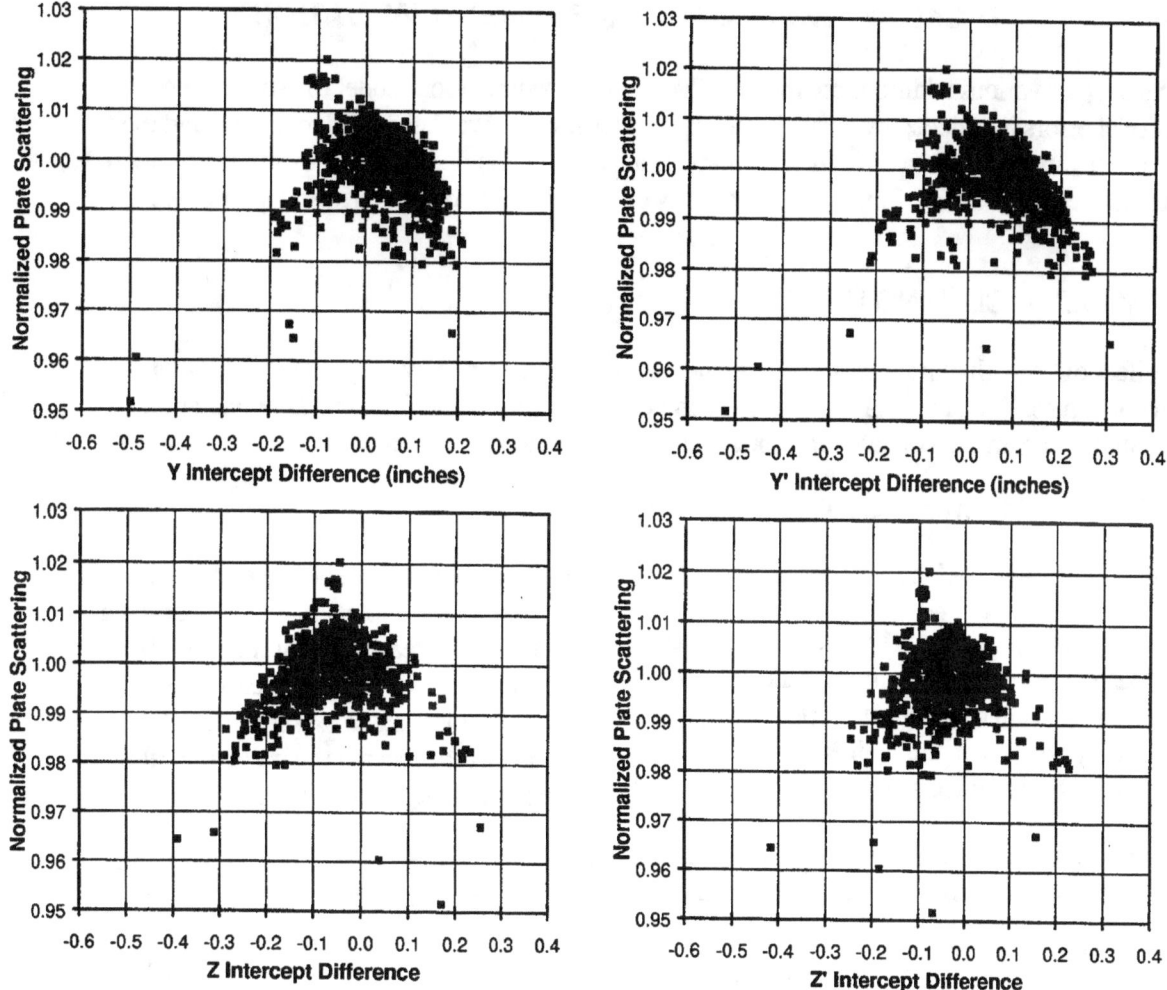

Figure 56. Dependence of Plate Scattering on Intercept Differences

Figure 57. Dependence of Plate Scattering on Intercept Differences Rotated by 27 Degrees

In Chapter 4, it will be shown that a 27 degree rotation in the Y-Z plane will separate intercepts that are in the scattering plane from those that are out of the scattering plane. Figure 57 shows the effect of the same 27 degree rotation on Figure 56. The plate scattering dependence noted in Figure 56 for both Y and Z components appears for only the Y' component in Figure 57. The rotated version of Figure 51 was relatively unaffected by the rotation and is not included. The analysis of Section 4.1.1.2 will provide an explanation of how to interpret the Y' dependence of Figure 57.

4. CALIBRATION MODEL IMPROVMENT/VALIDATION

In this chapter various additions are made to the calibration simulation model to estimate the additional errors introduced by errors in the transmitter and receiver heads and to understand the calibration variances observed in the last chapter. In addition, the field measurements are compared to the calculated calibrations. Finally, the suitability of the plate and fog scattering properties are examined.

4.1 MODEL INCLUDING HEADS

The effect of head alignment errors on the calculation was examined by varying the calibrator intercept values. The same analysis method can be used to understand the calibration variation associated with variation of calibrator intercept values described in Figure 50.

4.1.1 Large Receiver Head Alignment Errors

4.1.1.1 Unrotated Y-Z Coordinates

A fork with small calibrator intercept offsets (raw S/N 642) was selected for analyzing the impact of large changes in receiver intercept. Presumably changes in transmitter intercept would have similar effects.

Figures 58 and 59 show how the beam overlap is affected by large changes in receiver Y and Z intercept, respectively. The lower plot expands the plot for small displacements. Of particular interest is the plateau in the number of plate scattering points for displacements of less than 0.5 inches. This plateau exists because the projection of the receiver beam on the calibrator is larger than that of the transmitter.

The effect of intercept changes on plate and volume overlap is similar for Z coordinate

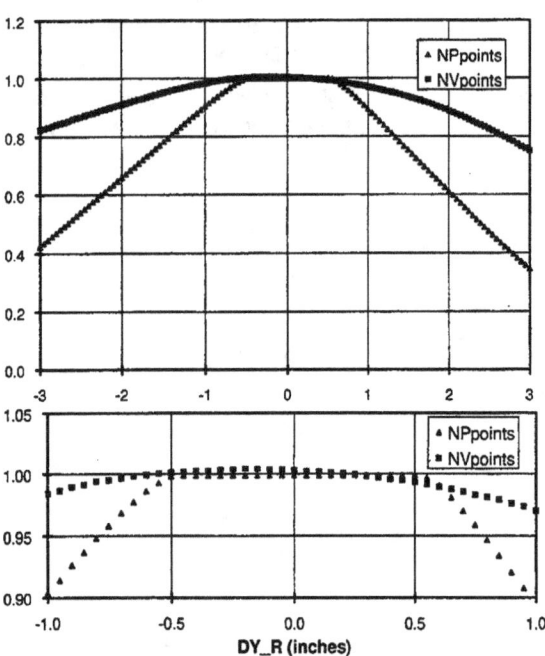

Figure 58. Dependence of Normalized Number of Points on Y Intercept Changes

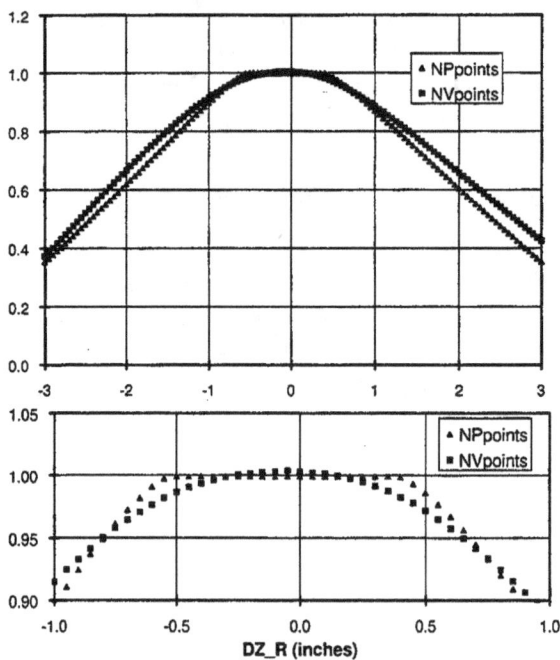

Figure 59. Dependence of Normalized Number of Points on Z Intercept Changes

changes, but different for Y coordinate changes. This difference will be explained in the next section.

Figures 60 and 61 show the effect of receiver intercept changes on normalized plate and volume scattering and NRVP (NRatVP in figures). As in Figures 58 and 59, the upper plot in each figure shows a wide variation and the lower plot shows a magnified view, that covers the area related to the calibration analysis of the last chapter.

4.1.1.2 Rotated Y-Z Coordinates

The relationships shown in Figures 59 through 61 are more complex than the results of earlier analyses because the changes in both Y and Z intercept coordinates vary both the overlap between the transmitter and receiver beams and mean scattering angle.

Figure 62 shows the variation in mean scattering angle with intercept change. The scattering angle variation is about twice as large for Y as for Z. This variations suggests that rotating the Y-Z coordinate system could isolate the scattering angle effect to the plane of the scattering, namely the new Y' coordinate. The proper angle of rotation has a tangent of 0.5 (27°), which is roughly the ratio of the out-of-plane angle (8-10°) to the tilt down angle (19°). Figure 63 shows the resulting dependence of mean scattering angle on intercept displacement. The mean scattering angle is almost constant for changes in Z'. All the variation has been shifted to the Y' coordinate.

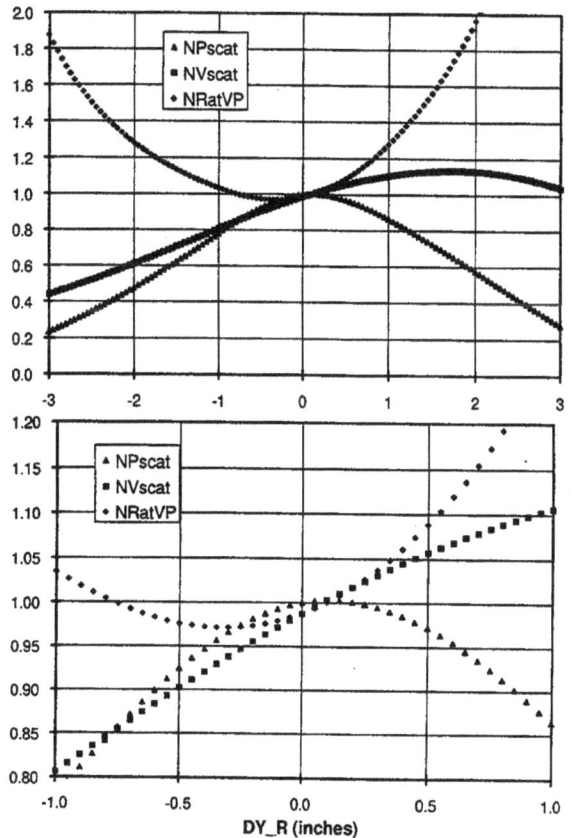

Figure 60. Effect of Y Intercept Changes

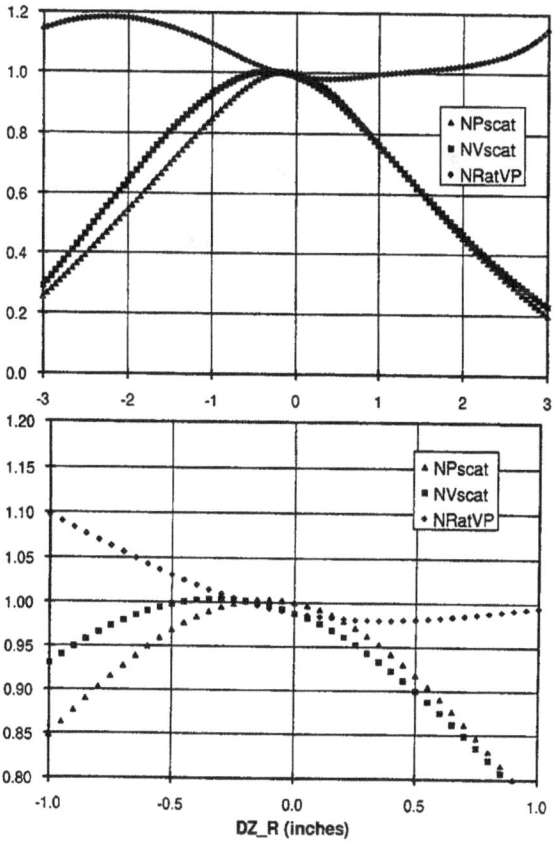

Figure 61. Effect of Z Intercept Changes

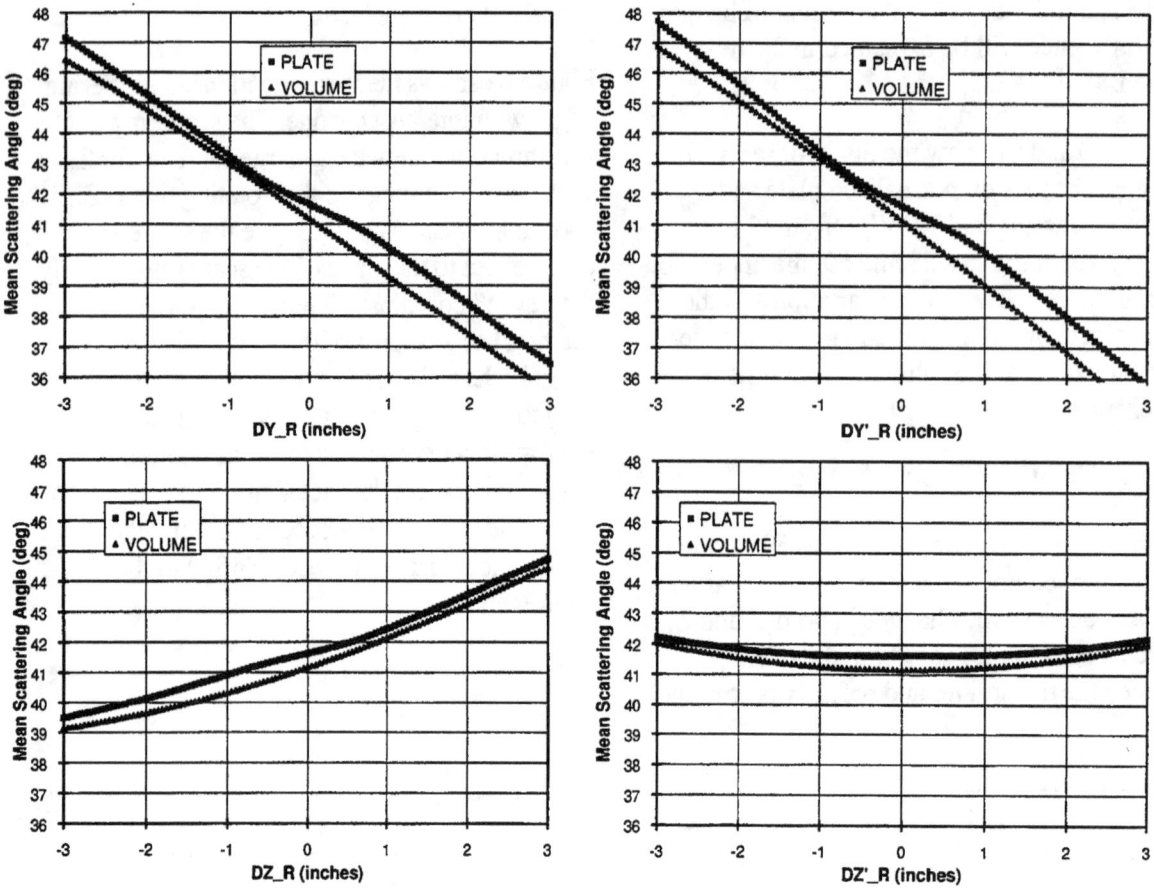

Figure 62. Variation of Mean Scattering Angle with Receiver Intercept Displacement

Figure 63. Mean Scattering Angle vs Receiver Intercept Displacement, after Rotation

Figures 64 and 65 show how the number of points varies with Y' and Z' displacement, respectively.

With the rotation, the Z' dependence of beam overlap becomes the same for both volume and plate scattering.

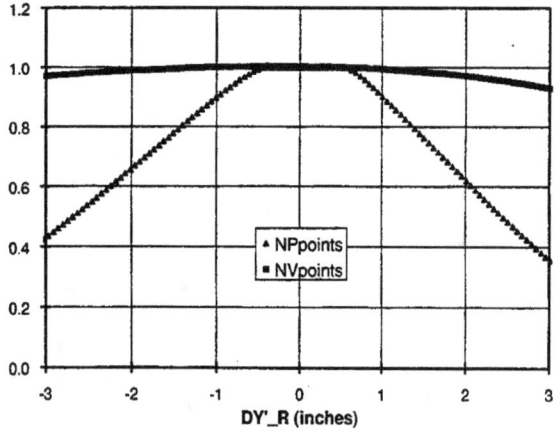

Figure 64. Points vs Y' Displacement

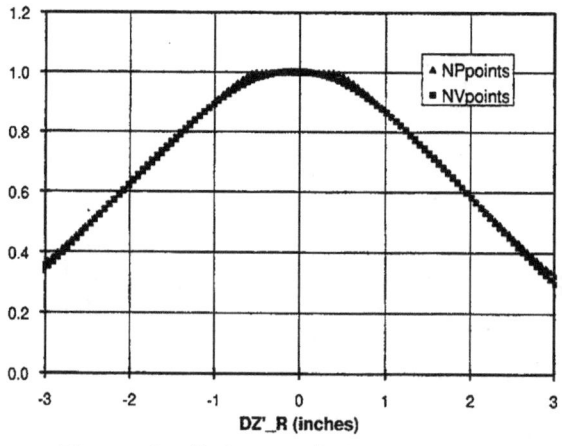

Figure 65. Points vs Z' Displacement

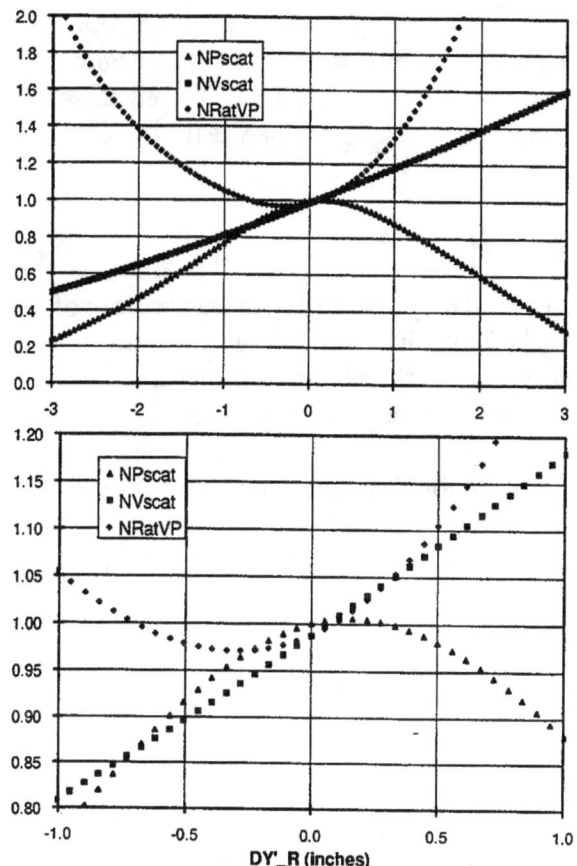

Figure 66. Effect of Y' Intercept Changes

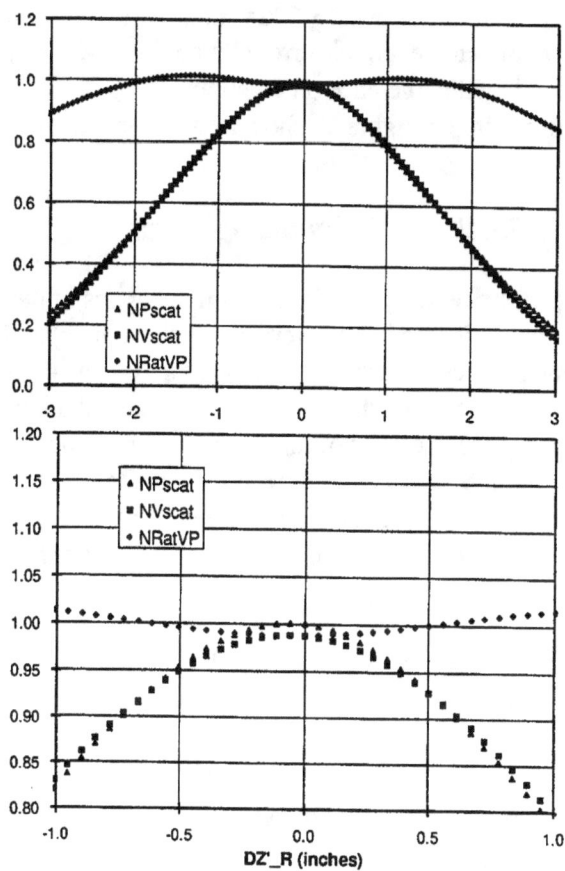

Figure 67. Effect of Z' Intercept Changes

Figures 66 and 67 show the scattering effects of displacements in the new coordinate system. The results are easier to understand than those in Figures 60 and 61.

The DZ' dependence (Figure 67) is primarily caused by overlap effects, which are almost identical (Figure 65) for plate and volume scattering. Consequently, the volume/plate scattering ratio is relatively unaffected by changes in DZ'.

The DY' dependence is more complex, because it combines both overlap effects (see Figure 64) and scattering angle effects (see Figure 63). The scattering angle decreases rapidly with increasing DY'. Decreasing scattering angle increases volume scattering (-0.07 per degree) more than twice as fast as plate scattering (-0.03 per degree).

Because the volume scattering overlap varies little with DY' in Figure 64, the volume scattering variation in Figure 66 is completely dominated by the scattering angle effect and leads to a continuous increase in volume scattering with DY'.

On the other hand, the overlap effect for plate scattering for DY' (Figure 65) is similar to that for DZ' (Figure 66). The net effect for plate scattering in Figure 67 is a combination of the overlap effect and the increased scattering with decreasing scattering angle, which leads to a plate scattering peak that is displaced toward negative DY'.

The resulting ratio of volume scattering to plate scattering in Figure 66 is dominated by the volume scattering effect and increases rapidly and linearly with DY'. At the limits of the intercept tolerance (± 0.30 inches), it would use up almost all the allowed ± 7% variation in calibration. However, this error source is bounded by the tolerance on compound scattering angle; Figure 59 shows that the mean volume scattering angle varies by two degrees per inch of Y' displacement. The ± 0.25 degree tolerance on compound angle would translate into limits on Y' displacement of ± 0.125 inches.

4.1.2 Small Head Alignment Errors

The process used by Handar to align the optics of the sensor head may give an alignment accuracy of about ± 0.2 degree in both vertical (Y coordinate) and lateral (Z coordinate) directions. The head mount gives essentially no tolerance for vertical alignment errors, but the alignment pins permit approximately ± 0.3 degree variations in lateral alignment. The lateral alignment tolerance might be less in practice because of the alignment properties of the two captured screws used to attach the head to the mount. In any case, the ± 0.2 degree tolerance will be taken for Y alignment and ± 0.3 degree tolerance will be taken for Z alignment. These head alignment tolerances correspond approximately to calibrator displacements of ± 0.10 inches for Y and ± 0.15 inches for Z. Four cases were analyzed:

1. DY_R = 0.10", DY_T = 0.10"
2. DY_R = 0.10", DY_T = -0.10"
3. DZ_R = 0.15", DZ_T = 0.15"
4. DZ_R = 0.15", DZ_T = -0.15"

The cases with same sign displacement for receiver and transmitter will give maximum scattering angle change and minimal beam overlap change. Conversely, the cases with opposite sign displacement for receiver and transmitter will give minimal scattering angle change and maximum beam overlap change.

Figure 68 shows the results of the analysis as a function of raw fork serial number. The top plot shows the overall resulting spread in the NRVP. The extent of the spread is only slightly greater than observed in Figure 22 for the undisplaced analysis. Only one NRVP value (for one early production unit) is outside the ± 7% error limits.

The bottom of Figure 68 shows how the displacements affected the calculated NRVP. The same sign displacements (scattering angle change, no overlap change) gave very consistent changes: +0.0013 inches for Y displacements and -0.009 inches for Z displacements. Even though the Z displacement was larger, it has less effect on NRVP than the Y displacement because it generated a smaller change in scattering angle. The opposite sign changes (no scattering angle change, significant overlap change) had generally smaller, but much more diverse effects on NRVP. Most of the cases where the overlap effects were greater than the scattering angle effects were early production forks.

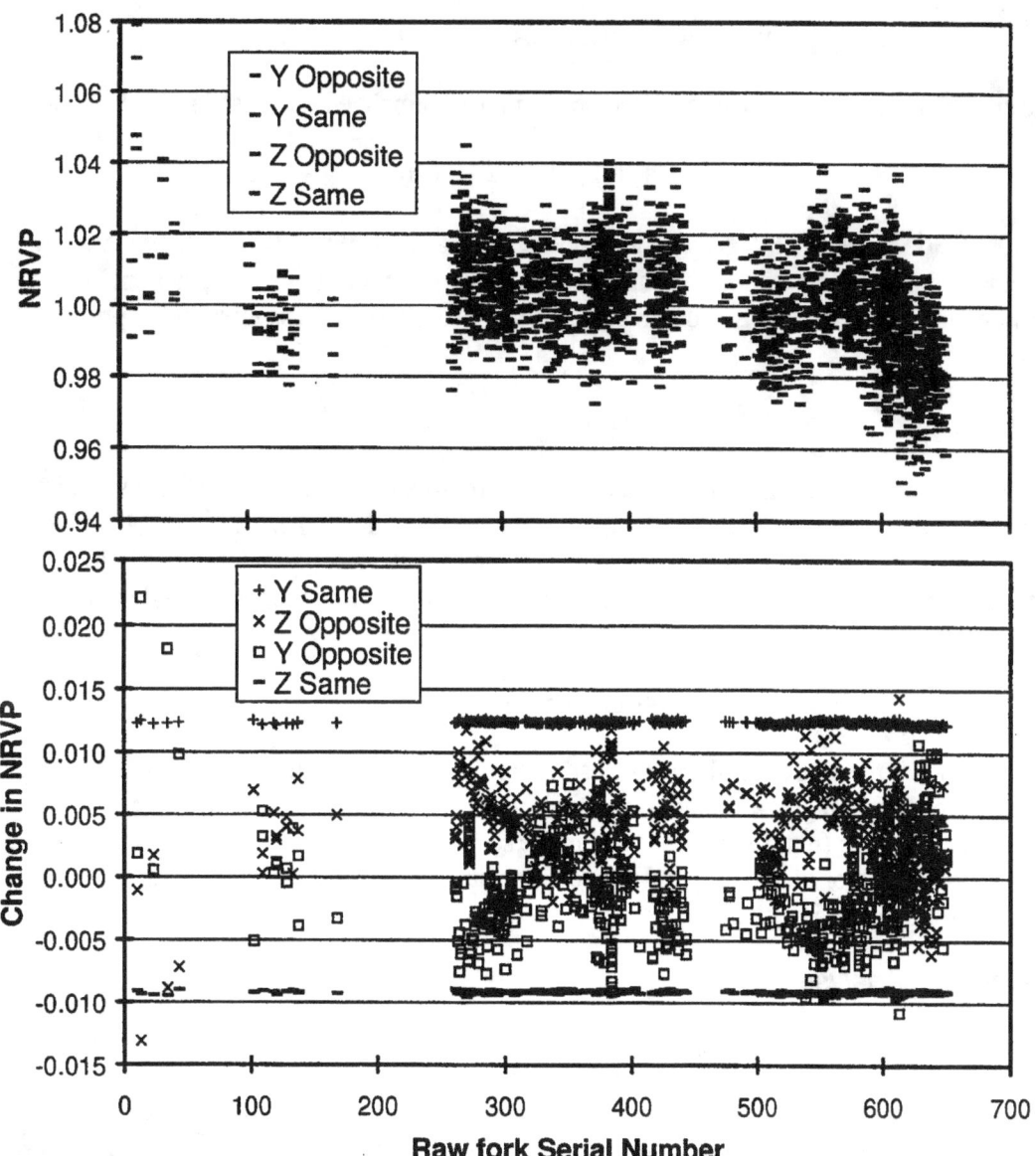

Figure 68. Effect of Intercept Displacements on NRVP (top) and Change in NRVP (bottom)

4.2 PLATE AND FOG SCATTERING PARAMETERS

4.2.1 Plate

If the plate is modeled as a perfect diffuse scatterer combined with the transmission holes, the scattered response will be proportional to:

$$\cos(\alpha 1)\cos(\alpha 2) \qquad (1)$$

where $\alpha 1$ and $\alpha 2$ are the angles with respect to the normal of the beams entering and leaving the calibration plate. If the angles are expanded around the nominal value $\alpha 0$, the scattered response becomes proportional to:

$$\cos(\alpha 0) \cos(\alpha 0) - \sin(\alpha 0) \cos(\alpha 0) (\alpha 1 + \alpha 2 - 2\alpha 0) \qquad (2)$$

where the angles are in radians. Equation 2 can be put in the normalized form used in the calculation:

$$1 - \tan(\alpha 0) (\alpha 1 + \alpha 2 - 2\alpha 0) \qquad (3)$$

Because $\alpha 0$ is 0.366 radians (21 degrees), the fractional change in response with change in radian scattering angle is $-\tan(\alpha 0) = -0.384$. The value needed for input to the simulation program becomes -0.0067 per degree, which is much smaller than the value -0.03 per degree assumed in the prior calculation. (2)

4.2.2 Fog

Although fog modeling has been conducted by a number of different groups, the angular variation of scattering, required by the calibration simulation, is not readily available. An available[7] plot of scatter function for cumulous clouds may give some guidance for what the fog dependence might look like. This plot shows an exponential relationship between scatter function and scattering angle Θ over the angle range 15 to 60 degrees:

$$\exp(\alpha \Theta), \alpha = -0.07 / \text{deg}. \qquad (4)$$

This relationship was the source for the original selection of -0.07 /deg to represent the linear in fog scattering with scattering angle. Although Equation 4 was calculated for clouds and optical wavelength = 0.7 microns (not 0.9 microns used in the scatterometer), it is consistent with the Otis test data which showed a factor of two drop in HSS scatterometer fog response when the scattering angle was increased from 35 to 45 degrees.

4.2.3 Sensitivity Analysis

Since the scattering properties assumptions are most closely associated with the influence of fork compound angle on the calculated calibration, two cases were selected from the ends of the normal variation of NRVP with compound angle in Figure 40. The compound scattering angles were 41.576 and 42.460 degrees for Cases 1 and 2, respectively.

Table 14. Results of Sensitivity Analysis

Plate b (/deg)	Volume b (/deg)	Volume Form	Case 1 NRVP	Case 2 NRVP
-0.03	-0.07	Lin	1.025	0.960
-0.03	-0.07	Exp	1.026	0.960
-0.007	-0.07	Lin	1.030	0.947
-0.007	-0.10	Lin	1.037	0.929

To normalize the results, the calculations were compared to the data from the Teledyne case with assembly S/N = 227 (Birmingham fork in Table 7 with NRPV = 1.000). Table 14 presents the results of this analysis:

1. The top line in Table 14 presents the baseline results. The calculated calibration error is +2.5% for Case 1 and -4.0% for Case 2.

2. The second line in Table 14 shows the effect of using an exponential variation in volume scattering with scattering angle. The amount of volume scattering was slightly reduced for all cases, but the NRVP values were little changed from the baseline values.

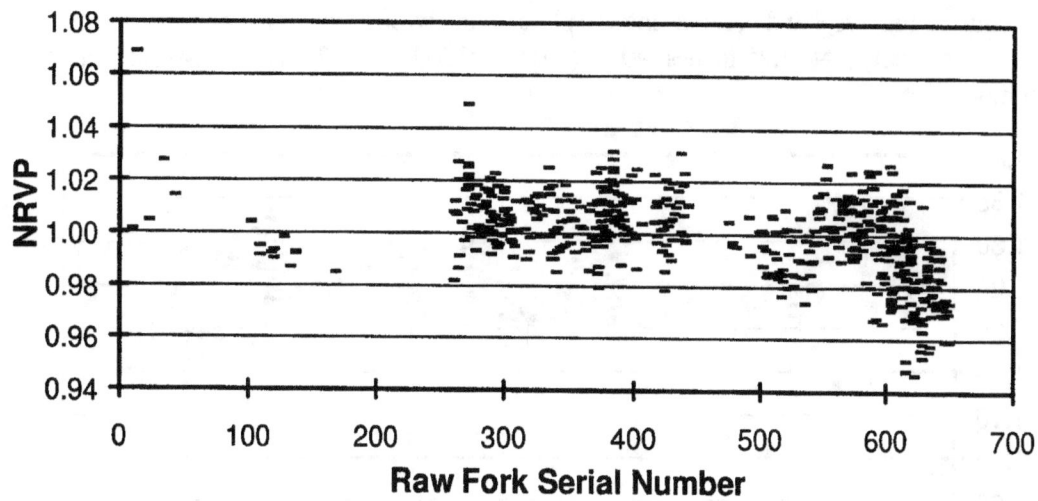

Figure 69. Calibration Distribution for Case 3

3. The third line in Table 14 shows the effect of reducing the plate dependence to the theoretical value derived in Section 4.3.1. This change increases the errors to 3.0 and -5.3 % for Cases 1 and 2, respectively. The increase in error is a factor of 1.2 and 1.3 for the two cases. The increase in error is only half of the increase in the difference between plate and volume scattering b values for the two analyses: (-0.063)/(-0.04) =1.6. Since this case represents the best estimate of both volume and plate scattering, it will be used in subsequent analyses in this chapter. Figure 69 shows the distribution of calibrations through the production run for this case. It is slightly broader than the baseline calculations in Figure 22. The mean RVP value for the 87 Teledyne forks is very slightly changed (from 5.66 to 5.65).

4. The fourth line in Table 14 shows the effect of increasing the volume dependence from -0.07 to -0.10, which is perhaps an upper limit to the possible value. This change increases the errors to 3.7 and -7.1% for Cases 1 and 2, respectively. The Case 2 value hits the error limit. The increase in error from line three to line four is a factor of 1.2 and 1.3 for the two cases. Again, the increase in error is roughly half of the increase in the difference between plate and volume scattering b values for the line three and line four analyses: (-0.093)/(-0.063) =1.5.

The reduced plate scattering in line three of Table 14 appears to be well justified by the analysis of Section 4.3.1. The analysis of Section 4.3.2 gives reasonable justification for the baseline value for volume scattering. Thus, the sensitivity analysis would suggest a factor of 1.2 to 1.3 increase in error, which will have relatively little impact on the accuracy for most forks.

4.3 WRONG CALIBRATOR LOCATION

When the scattermeter scattering geometry was changed from look-across to look-down, the calibrator was moved to the opposite side of its mount. Figure 10 shows the correct mounting configuration for the look-down forks. The thumbscrew at the bottom is on the opposite side for the look-across fork. Because both mounting arm configurations were available for many years after the change from look-across to look-down geometry (the arm with calibrator S/N 0001 is still the look-across configuration), it is possible for incorrect calibrations to result if the wrong arm was used. The simulation model can be used to calculate the impact of using the wrong arm configuration. As can be seen by examining Figures 8 and 9, mounting the plate on the wrong side of the calibrator arm will place it at approximately x = +1.25 inches = 0.99 (mount thickness) +0.26 (twice displacement of plate from mount).

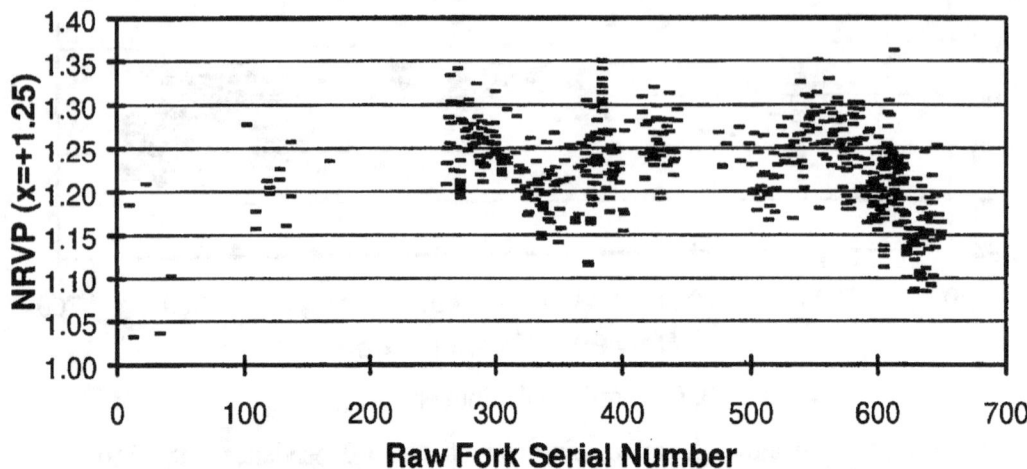

Figure 70. Normalized Scattering Ratio for Displaced Calibrator

Figure 70 shows the calculated calibrations for the incorrectly located calibrator plate. The NRVP value is typically 15 to 30 % high and the spread is much greater than for the correct calibrator location.

4.4 FIELD TESTED EARLY PRODUCTION UNITS

Because the calculated ratio of volume to plate scattering varied by more than 1% for only two of the scattermeters deployed for field tests, the opportunity for using field test data to validate the calibration simulation is very limited. The usable sensors are two preliminary Otis sensors in Table 3 (TDN1 with RPV = 0.978 and TDN2 with RPV = 0.964). The field test data for these sensors (Figures 4 and 5) suggest MOR ratios of roughly 1.00 to 1.02 for both sensors. If the normal MOR ratio for Sensors with RPV = 1.00 is taken as 0.94, then the predicted values of MOR ratio for TDN1 and TDN2 would be 0.94 and 0.93, which is far from the measured values. In fact, the model predicts the opposite calibration deviation from the normal sensors from what is observed. However, the measurements of the early production sensors giving calculated RPV close to 1.00 also show measured median MOR ratios usually greater than 1.00. Thus the discrepancy might be related to the time of testing rather than the calibration modeling. One possible source for the discrepancy was the early confusion about which side of the mounting post was to hold the calibration plate. The correct side was changed when the scattering geometry was changed from look-out to look-down, but not all calibrators were corrected immediately. This error corresponds to a calibrator location x displacement of about 1.3 inches, which is larger than the 0.8 inches error of the SIMCAL8 program.

5. CONCLUSIONS

5.1 CALIBRATION MODELING

5.1.1 SIMCAL8 Validation

Careful analysis of the code and output from the SIMCAL8 code uncovered two errors. When the errors were corrected, the calculated variance in calibration with fork geometry was reduced dramatically. The resulting calibration error distribution leaves substantially more tolerance for other error sources within the ± 7% unit-to-unit allowed variation in fog response.

5.1.2 Sensitivity to Parameters

The worst case considered by the sensitivity analysis would push the calibration of one end of the normal fork distribution to the 7% error limit. The most likely change in scattering parameters would lead to a factor of 1.2 to 1.3 increase in the calculated errors. Such an increase would have little impact on accuracy for most forks.

5.1.3 Comparison to Field Tests

The field test results for the two forks with calculated calibrations far from nominal were not consistent with the calculations. The problem might be related to the early test issue of which side of the calibrator post should have the plate mounted.

5.1.4 Effect of Head Variations

The original calibration simulation considered only variations in fork geometry. Adding reasonable head alignment errors changed the calibration of most forks by only 2.2% for the worst possible combination of errors (both Z and Y displacements of both heads).

5.2 PRODUCTION CONSISTENCY

5.2.1 Forks

Most the forks measured had consistent geometries. The two exceptions are the early production forks that showed more variance and the last 30 forks that showed a systematic shift in geometry.

5.2.2 Effect of Heads

Adding head alignment errors to the fork errors resulted in relatively minor broadening of the fork calibration spread.

6. RECOMMENDATIONS

The field test results were inconclusive (Section 4.4) for validating the simulation model. Resolving the observed inconsistencies will require additional testing of forks far from the golden geometry that was used for must field testing to date. Testing at Otis should be conducted on several forks with abnormal geometry. For example, fork AS/N = 360 (RVP=1.057) could be sent to Otis for additional testing.

Bent forks from early production should be measured with the CMM machine and retired from service if the calculated calibration is outside the allowed ± 7% range.

More detailed production information should be obtained from Teledyne or Handar.

CMM measurements should be obtained for future fork procurements.

A definite plan should be developed for long-term calibration stability.

APPENDIX A - CMM PROGRAM

A.1 PROGRAM

Annotations have been added to the program to identify what is being measured.

A.1.1 Original

```
 1: PROGRAM H : FORK 1E)
  2; SETUP 2100 3.0000
  3: TITLE / DEF "###################################################" @
     + "##############################" "###################################################" @
     + "##################################################"# @
     + "    HANDAR INC                                                            #" "#  " @
     + "                     1288 REAMWOOD AVE.                                        " @
     + "                                                  SUNNYVALE, CA 94089         " @
     + "                                                                 TEL- (408" @
     + ") 734-9646                                                 #" "#              " @
     + "          FAX. (408) 745-7921                                                 "
  4: INQUIRIES / DEF "SERIAL NUMBER" "PART NUMBER" "REV. LETTER" "PART NAME" @
     "COMMENTS" "COMENTS" "COMMENTS" "COMMENTS "
  5: PARAMETER / UNIT INCH
  6: PARAMETER / ANGLE DEG_MIN_SEC
  7: PARAMETER / SCALE 1.0000
  8: PARAMETER / DECIMAL 4
  9: PARAMETER / COORDINATES CARTESIAN
 10: PARAMETER / PROBE_POSITION X = 0.0000 Y = 0.0000 Z = 0.0000
 11: DEF_MASTERBALL / NUMBER 1 X = 22.9550 Y = 29.9558 Z = 6.6169 @
     DIAMETER = 1.0000
 12: STATISTICS / OFF
 13: BEGIN
 14: PRINT INFO / MESSAGE SCREEN N_PROMPT "CAUTION!!!!  THIS IS A CNC PROGR" @
     + "AM, PLEASE REMOVE" "ALL OBSTACLES FROM THE TABLE, AFTER YOU HAVE"
        "PRESSED ENTER THE MACHINE WILL MOVE TO THE HOME POSITION."
 15: PAUSE /
 16: CNC_MODE / ON
 17: GOTO / HOME
 18: PRINT INFO MESSAGE SCREEN NO_PROMPT "PLACE THE PART IN THE FIXTURE A" @
     + "ND PRESS ENTER," "THE CNC MEASUREMENT PROGRAM WILL BEGIN."
 19: PAUSE /
 20: PRINT_INFO / MESSAGE SCREEN NO_PROMPT "PLACE THE PART IN THE FIXTURE, " @
     + "PRESS ENTER AND" "THE CNC MEASUREMENT PROGRAM WILL BEGIN."
 21: GOTO / CURRENT CARTESIAN X = 6.1708 Y = 42.6,340 Z = 66705
 22: CHANGE_STYLUS / PH9 4
 23: CNC_FACTOR / SAFETY 4.0000
Start saving each element in file
 24: STORE_ASCII / ON

Element A1, Plane, top plane of receiver mount, measured as three points
 25: PLANE / MEASURE NUMBER 3
 26: GO_MEASURE / CURRENT CARTESIAN X = 2.8259 Y = 45.7890 Z = 40493 @
     SPATIAL L = 135.0559 M = 45.0559 N = 90.0000
 27: GO_MEASURE / CURRENT CARTESIAN X = 2.0051 Y = 43.4834 Z = 3.4013 @
     SPATIAL L = 150.5937 M = 60.5937 N = 90.0000
 28: GO_MEASURE CURRENT CARTESIAN X = 2.0167 Y = 43.4832 Z = 4.4217 @
     SPATIAL L = 142.0907 M = 52.0907 N = 90.0000
```

Define top receiver mount plane as x-y plane
  29: BASE_PLANE / XY PLANE LAST_ELEMENT

Element A2, Line, long side of receiver mount, measured as two points, ¼ inch below x-y plane
  30: LINE / MEASURE NUMBER 2
  31: GOTO / CURRENT CARTESIAN X = -1.1063 Y = -0.8540 Z = 0.1795
  32: GO_MEASURE / CURRENT CARTESIAN X =-0.7995 Y = -1.2277 Z = -0.2500
      SPATIAL L = 0.3923 M = 89.4658 N = 89.2251
  33: GO_MEASURE / CURRENT CARTESIAN X = -1.2383 Y = 1.4099 Z = =0.2500
      SPATIAL L = 0.3923 M = 89.4658 N = 89.2251
Align y axis with long side of receiver mount
  34: ALIGN / AXIS Y_AXIS LAST ELEMENT
  35: GOTO / CURRENT CARTESIAN X = -1.2640 Y = 2.5120 Z = -0.200

Element A3, Line, short side of receiver mount, measured as two points
  36: LINE / MEASURE NUMBER 2
  37: GO_MEASURE / CURRENT CARTESIAN X = -0.7420 Y = 2.2127 Z = -0.2500
      SPATIAL L =98.5913 M = 166.4058 N = 90.0000
  38: GO_MEASURE / CURRENT CARTESIAN X = 0.6262 Y = 2.2128 Z = -0.2500
      SPATIAL L = 98.5422 M = 158.3326 N = 90.0000

Element A4, Point, Corner of receiver mount, as intersection two sides
  39: POINT / INTERSECTION LAST_ELEMENTS
Make corner origin of coordinate system
  40: ALIGN / ORIGIN XYZ LAST_ELEMENT
  41: CNC / SAFETY 2.0000

Element A5, Plane, top surface of receiver mount as best fit to 8 points
  42: PLANE / MEASURE NUMBER 8
  44: GO_MEASURE / CURRENT CARTESIAN X = 1.3916 Y = -0.1803 Z = 0.0786
      SPATIAL L = 87.5358 M = 80.5033 N = 180.0000
  45: GO_MEASURE / CURRENT CARTESIAN X = 0.6036 Y = -0.2159 Z = 0.0803
      SPATIAL L = 88.4355 M = 86.0237 N = 180.0000
  46: GO_MEASURE / CURRENT CARTESIAN X = 0.4034 Y = -0.8140 Z = 0.0785
      SPATIAL L = 87.0207 M = 75.1939 N = 180.0000
  47: GO_MEASURE / CURRENT CARTESIAN X = 1.7251 Y = -0.7967 Z = 0.0765
      SPATIAL L = 69.1453 M = 91.0058 N = 180.0000
  48: GO_MEASURE / CURRENT CARTESIAN X = 1.4647 Y = -2.4116 Z = 0.0760
      SPATIAL L = 87.3742 M = 79.0744 N = 180.0000
  49: GO_MEASURE / CURRENT CARTESIAN X = 0.4414 Y = -2.4170 Z = 0.0771
      SPATIAL L = 89.2956 M = 90.4641 N = 180.0000
  50: GO_MEASURE / CURRENT CARTESIAN X = 0.4826 Y = -3.2511 Z = 0.0807
      SPATIAL L = 88.1910 M = 83.2841 N = 180.0000
  51: GO_MEASURE / CURRENT CARTESIAN X = 1.4181 Y = -3.2762 Z = 0.0802
      SPATIAL L = 86.5643 M = 74.4439 N = 180.0000
Redefine x-y plane as 8-point Top Receiver Mount
  52: BASE_PLANE / XY_PLANE LAST_ELEMENT
Print tolerance elements only
  53: PRINT-ELEMENT / ON TOLERANCE_ONLY
Print title
  54: PRINT_ELEMENT / TITLE
  55: HEADING / ONE_HEADING_PER_PAGE
  56: PRINT_INFO / MESSAGE PRINTER NO_PROMPT "FLATNESS OF PLANE ON LONG FORK"
      + "."
  57: FORM_TOL / FLATNESS FORM = 0.0100 NO_TRANSFER

Element A6, Circle, 1st receiver alignment pin, measured as 4 points

```
 58: CIRCLE / MEASURE NUMBER 4
 59: AUTO.-MEASURE_CIRCLE / CURRENT OUT NUMBER 4 CENTER CARTESIAN X = @
     -0.6908 Y = -1.8780 Z = 0.1500 DIAMETER = 0.2500 XY_PLANE POSITIVE @
     START_ANGLE = -90.000 TOTAL_ANGLE - 360.0000 MEASURE_DIRECTION @
     CLOCKWISE DEPTH = 0.0000 AVOID = 0.3500 LINEAR_MOVE
```

Element A7, Circle, 2ND receiver alignment pin, measured as 4 points
```
 60: CIRCLE / MEASURE NUMBER 4
 61: AUTO_MEASURE_CIRCLE / CURRENT OUT NUMBER 4 CENTER CARTESIAN X = @)
-0.7110 Y = 1.4290 Z = 0.1500 DIAMETER = 0.2500 XY_PLANE POSITIVE @
     START_ANGLE = -90.000 TOTAL_ANGLE - 360.0000 MEASURE_DIRECTION @
     CLOCKWISE DEPTH = 0.0000 AVOID = 0.3500 LINEAR_MOVE
 62: PRINT_INFO / MESSAGE PRINTER NO_PROMPT "DISTANCE BETWEEN PINS ON LONG " @
     + "FORK."
```

Element A8, Distance, distance between receiver alignment pins
```
 63: DISTANCE / LAST_ELEMENTS
 64: TOLERANCE / DISTANCE NOMINAL = 3.5850 UPPER_TOL = 0.0060 LOWER_TOL = @
     -0.0060 NO_TRANSFER
```
Move x-y origin to element A7 (7th element measured, i.e., 2nd alignment pin at line 59)
```
 65: ALIGN / ORIGIN XY A7
```
The following operation rotates the coordinate system around the z axis (at A7) by an angle which puts A6 at an angle corresponding to x = -1.4 and y = -3.3, the nominal design value. This defines the nominal beam direction and puts the y-axis down the receiver beam center
```
 66: ALIGN / AXIS_OFFSET POINT A6 CARTESIAN DIST_FROM_FIRST_AXIS = @
     -3.3000 DIST_FROM_SECOND_AXIS = -1.4000
```

Element A9, Point, bisector, center of pins in receiver mount
```
 67: POINT / BISECTOR A6 A7 CENTER
```
Move x-axis origin to midway between pins
```
 68: ALIGN / ORIGIN X LAST_ELEMENT
```
On y axis, move origin to hole position from alignment pin position.
```
 69: ALIGN / ORIGIN OFFSET X = 0.0000 Y = -1.3500
```

Element A10, Point, Location of hole in receiver mount in top plane
```
 70: POINT / KEYIN CARTESIAN X = 0.0000 Y     0.0000 Z= 0.0000 NO_PROMPT
```
Save Location of hole in receiver mount in top plane
```
 71: MEMORY / M1
```

Element A11, Line, receiver beam line at hole location
```
 72: LINE / KEYIN CARTESIAN X = 0.0000 Y = 0.0000 Z = 0.0000 SPHERICAL @
     PHI = -90.0000 THETA = 90.0000 NO_PROMPT
```
Save receiver beam line
```
 73: MEMORY / M2
 74: GOTO / CURRENT CARTESIAN X = 0.5260 Y = 0.0896 Z = 0.6134
 75: GOTO / CURRENT CARTESIAN X = -5.4834 Y = -36,4845 Z = 13.6072
 76: CHANGE_STYLUS / PH9 3
 77: CNC_FACTOR / SAFETY 4.0000
 78: CNC-MODE / OFF
```
Go to manual mode to make sure can find transmitter mount
```
 79: PRINT_INFO / MESSAGE SCREEN NO_PROMPT "MANUALLY MEASURE 3 POINTS"
```

Element A12, Plane, top of transmitter mount
```
 80: PLANE / MEASURE NUMBER 3
```
Assign as x-y plane
```
 81: BASE PLANE / XY_PLANE LAST_ELEMENT
 82: PRINT_INFO / MESSAGE SCREEN NO_PROMPT "CAUTION CNC MODE WILL START"
```

```
 83: PAUSE /
 84: CNC_MODE / ON
 85: GOTO / CURRENT CARTESIAN X = -1.7754 Y = -1.2676 Z = 0.1888

Element A13, Line, long side of transmitter mount
 86: LINE / MEASURE NUMBER 2
 87: GO_MEASURE / CURRENT CARTESIAN X = -1.4903 Y = -1.1366 Z = -0.2550 @
     SPATIAL L = 11.0202 M = 101.0109 N = 90.0000
 88: GO_MEASURE / CURRENT CARTESIAN X = -0.5547 Y = 1.3849 Z = -0.2550 @
     SPATIAL L = 11.1111 M = 101.0923 N - 90.0000
Align y-axis to long side of mount
 89: ALIGN / AXIS Y_AXIS LAST_ELEMENT
 90: GOTO / CURRENT CARTESIAN X = -1.3295 Y = 2.0976 Z = -0.2528

Element A14, Line, short side of transmitter mount
 91: LINE / MEASURE NUMBER 2
 92: GO_MEASURE / CURRENT CARTESIAN X = -0.4102 Y = 1.8955 Z = -0.2445 @
     SPATIAL, L = 80.3706 M = 161.0104 N = 90.0000
 93: GO_MEASURE / CURRENT CARTESIAN X = 0.6474 Y = 1.8967 Z = -0.2445 @
     SPATIAL L = 80.4217 M = 162.5956 N = 90.0000

Element A15, Point, corner of transmitter mount
 94: POINT / INTERSECTION LAST ELEMENTS
Put x-y origin on corner
 95: ALIGN / ORIGIN XY LAST_ELEMENT
 96: GOTO / CURRENT CARTESIAN X = 1.5433 Y = 0.3134 Z = 0.6145
 97: CNC-FACTOR / SAFETY 2.0000

Element A16, Plane, top of transmitter mount, 8-point fit
 98: PLANE / MEASURE NUMBER 8
 99: GO_MEASURE / CURRENT CARTESIAN X = 1.4661 Y = 0.4796 Z = 0.0778 @
     SPATIAL L = 90.0424 M = 94.0552 N = 180.0000
100: GO_MEASURE / CURRENT CARTESIAN X = 0.4227 Y = -0.6900 Z = 0.0767 @
     SPATIAL L = 88.4715 M = 101.5126 N = 180.0000
101: GO_MEASURE / CURRENT CARTESIAN X = 0-5017 Y = -1.1670 Z = 0.0777 @
     SPATIAL L = 90.3604 M = 92.1234 N = 180-0000
102: GO_MEASURE / CURRENT CARTESIAN X = 1.5771 Y = -1.0425 Z = 0.0801 @
     SPATIAL L = 89.3818 M = 96.4303 N = 180.0000
103: GO_MEASURE / CURRENT CARTESIAN X = 1-5120 Y = -2.9954 Z = 0.0798 @
     SPATIAL L = 89.5226 M = 95.1742 N 180.0000
104: GO_MEASURE / CURRENT CARTESIAN X = 0.3342 Y = -3.1772 Z = 0.0799 @
     SPATIAL L = 91.0126 M = 87.0343 N = 180.0000
105: GO_MEASURE / CURRENT CARTESIAN X = 0.7647 Y = -3.5696 Z= 0.0783 @
     SPATIAL L = 88.5425 M = 101.0804 N = 180.0000
106: GO_MEASURE / CURRENT CARTESIAN X = 1.3368 Y = -3.4979 Z = 0.0786 @
     SPATIAL L = 90.0330 M = 94.1117 N = 180.0000
107: PRINT_INFO / MESSAGE PRINTER NO_PROMPT "FLATNESS OF PLANE ON SHORT FOR" @
     + "K.
108: FORM_TOL / FLATNESS FORM = 0.0100 NO_TRANSFER

Element A17, Circle, 1st alignment pin of transmitter mount
109: CIRCLE / MEASURE NUMBER 4
110: AUTO_MEASURE_CIRCLE / CURRENT OUT NUMBER 4 CENTER CARTESIAN X = @
     1.7125 Y = -3.5374 Z = 0.1500 DIAMETER = 0.2500 XY_PLANE POSITIVE @
     START_ANGLE = -90.0000 TOTAL-ANGLE = 360.0000 MEASURE_DIRECTION @
     CLOCKWISE DEPTH = 0.0000 AVOID = 0.3500 LINEAR_MOVE

Element A18, Circle, 2nd alignment pin of transmitter mount
```

```
111: CIRCLE / MEASURE NUMBER 4
112: AUTO_MEASURE_CIRCLE / CURRENT OUT NUMBER 4 CENTER CARTESIAN X = @
     0.2970 Y = -0.2490 Z = 0.1500 DIAMETER = 0.2500 XY_PLANE POSITIVE @
     START_ANGLE = -90.0000 TOTAL-ANGLE = 360.0000 MEASURE_DIRECTION @
     CLOCKWISE DEPTH = 0.0000 AVOID = 0.3500 LINEAR_MOVE
113: PRINT_INFO / MESSAGE PRINTER NO_PROMPT "DISTANCE BETWEEN PINS ON SHORT" @
     + " FORK."
```

Element A19, Distance, between alignment pins
```
114: DISTANCE / LAST_ELEMENTS
115: TOLERANCE / DISTANCE NOMINAL = 3.5850 UPPER_TOL = 0.0060 LOWER_TOL = @
     -0.0060 NO TRANSFER
```
Align x-y origin with 1st alignment pin
```
116: ALIGN / ORIGIN XY A17
117: ALIGN / AXIS_OFFSET POINT A18 CARTESIAN DIST_FROM_FIRST_AXIS = @
     3.3000 DIST_FROM_SECOND_AXIS = -1.4000
```

Element A20, Point, midpoint between transmitter alignment pins
```
118: POINT / BISECTOR LAST_ELEMENTS
```
Move x-origin to bisector
```
119: ALIGN / ORIGIN X LAST_ELEMENT
```
Move origin along y-axis to center of hole
```
120: ALIGN / ORIGIN_OFFSET X = 0.0000 Y = 1.3500 = 0.0000
```

Element A21, Point, center of transmitter hole in plane of mount top
```
121: POINT / KEYIN CARTESIAN X = 0.0000 Y = 0.0000 Z = 0.0000 NO_PROMPT
```
Save center of hole in transmitter mount as M3
```
122: MEMORY / M3
```

Element A22, Line, transmitter beam line, located at mount hole
```
123: LINE / KEYIN CARTESIAN X = 0.0000 y = 0.0000 Z = 0.0000 SPHERICAL @
     PHI = 90.0000 THETA = 90.0000 NO_PROMPT
```
Save transmitter mount perpendicular line as M4
```
124: MEMORY / M4
```

Element A23, Angle, compound scattering angle, supplement of angle between transmitter and receiver beams
```
125: ANGLE / M2 M4 SUPPLEMENT
126: PRINT_INFO / MESSAGE PRINTER NO_PROMPT "COMPOUND ANGLE BETWEEN PLATE."
127: TOLERANCE / ANGLE NOMINAL = 42.0000 UPPER_TOL = 0.1500 LOWER_TOL = @
     -0.1500 NO_TRANSFER
```

Element A24, Line, intersection of receiver top plane and transmitter top plane
```
128: LINE / INTERSECTION A5 A16
```
Define coordinate system with respect to this plane intersection, which is approximately perpendicular to plane of fork pipe.
```
129: BASE_PLANE / XY PLANE LAST ELEMENT
```
Get transmitter beam pointer
```
130: RECALL M4
```

Element A25, Line, rotated transmitter beam pointer
```
131: ROTATE / Z_AXIS ANGLE = -19.0000
```
Reverse original transmitter beam pointer
```
132: CHANGE_ELEMENT / DIRECTION A22
```
Recall rotated version of A22, stored as M4
```
133: RECALL / A25
```
Recall receiver beam pointer
```
134: RECALL./ A11
```

```
135: PRINT_INFO / MESSAGE PRINTER NO_PROMPT "PROJECTED ANGLE BETWEEN THE PL" @
     + "ATES."

Element A26, Angle, in-fork-plane angle between beams
136: ANGLE / LAST_ELEMENTS VECTOR_DIRECTION
137: TOLERANCE / ANGLE NOMINAL = 38.0000 UPPER_TOL = 0.3000 LOWER_TOL = @
     -0.3000 NO_TRANSFER
138: GOTO / CURRENT CARTESIAN X = 6.5946 Y = -13.0681 Z = 0.7233
139: GOTO / CURRENT CARTESIAN X = 1.6298 Y = -10.9951 Z = 5.5167
140: CHANGE_STYLUS / PH9 2
141: CNC_FACTOR / SAFETY 3.0000

Element A27, Plane, calibrator fixture plane
142: PLANE / MEASURE NUMBER 4
143: GO_MEASURE / CURRENT CARTESIAN X = 2.1829 Y = -0.1744 Z = 1.5388 @
     SPATIAL L = 90.2545 M = 1-3343 N = 91.3007
144: GO_MEASURE / CURRENT CARTESIAN X = 2.1886 Y = -0.1943 Z = 0.7000 @
     SPATIAL L = 88.3457 M = 1.5543 N = 91.1827
145: GO_MEASURE / CURRENT CARTESIAN X = 1.1518 Y = -0.2186 Z = 0.7000 @
     SPATIAL L = 90.2545 M = 1.3343 N = 91.3007
146: GO_MEASURE / CURRENT CARTESIAN X = 1.1470 Y = -0.2011 Z = 1.4365 @
     SPATIAL L = 90.2545 M = 1.3343 N = 91.3007

Element A28, y-axis intersection with calibrator plane A27, using dual-mount-plane
coordinate system of L129
147: POINT / INTERSECTION Y_AXIS A27
Move y-axis origin to calibrator plane
148: ALIGN / ORIGIN Y LAST_ELEMENT
Move to true calibrator location
149: ALIGN / ORIGIN OFFSET X = 0.0000 Y = -0.1300 Z = 0.0000

Element A29, Point, intersection of receiver beam with calibrator
150: POINT / INTERSECTION ZX_PLANE M2

Element A30, Point, intersection of transmitter beam with calibrator
151: POINT / INTERSECTION ZX_PLANE M4
152: PRINT_INFO / MESSAGE PRINTER NO_PROMPT "THIS IS THE DISTANCE BETWEEN T" @
     + "HE' INTERSECTION" "POINTS. "
153: PRINT_ELEMENT / ON ALL_ELEMENTS

Element A31, Distance, between beam intercepts at calibrator
154: DISTANCE / LAST_ELEMENTS
155: PRINT_ELEMENT / ON TOLERANCE_ONLY
156: GOTO / CURRENT CARTESIAN X = 1.0873 Y = 0.3466 Z = 13.5729
157: CHANGE_STYLUS / PH9 1

Element A32, Line, normal to calibrator at origin
158: LINE /KEYIN CARTESIAN X =0-0000 Y =0.0000 Z = 0 0000 SPHERICAL @
     PHI = 0.0000 THETA = 0.0000 NO_PROMPT
Save as M7
159: MEMORY / M7

Element A33, line between transmitter and receiver mount hole centers
160: LINE / CONNECT M3 M1
Change coordinate system to reference this line as x axis
161: BASE_PLANE / YZ_PLANE LAST_ELEMENT
Rotate around z axis by 90 degrees
162: ROTATE / Z_AXIS ANGLE = -90.0000
```

Element A34, Line, intersection of calibrator plane and plane perpendicular to head mount holes
163: LINE / INTERSECTION XY_PLANE A27
164: PROJECTION / ON YZ_PLANE
165: HEADING / ONE_HEADING_PER_PAGE
166: PRINT_INFO / MESSAGE PRINTER NO_PROMPT."" "PROJECTED ANGLE BETWEEN THE" @
    + " LONG FORK AND THE" "CENTERLINE."
167: RECALL / M2

Element A35, Line, reversed receiver beam line
168: CHANGE_ELEMENT / DIRECTION LAST_ELEMENT
169: TOLERANCE / PHI NOMINAL = 8.4200 UPPER-TOL = 1.0000 LOWER_TOL = @
    -1.000 NO TRANSFER
170: PRINT_INFO / MESSAGE PRINTER NO_ PROMPT "" "PROJECTED ANGLE BETWEEN THE" @
    + " SHORT FORK AND THE" "CENTERLINE-"
171: RECALL / M4
172: TOLERANCE / PHI NOMINAL = -10.3000 UPPER_TOL = 1.0000 LOWER_TOL = @
    -1.0000 NO_TRANSFER
173: GOTO / HOME
174: print_info / message both no_prompt "_____page 2 of 2_____"
175: print_info / message both no_prompt "_____END OF TEST_____"

A.1.2 Added July 1995

Element A36, Plane, Reversed direction calibrator plane
176: CHANGE_ELEMENT / DIRECTION A27
Use to define coordinate system relative to calibrator, as used in simulation
177: BASE_PLANE / YZ_PLANE LAST_ELEMENT
Put x origin at correct calibrator location, offset from calibration fixture
178: ALIGN / ORIGIN-OFFSET X = -0.1300 Y = 0.0000 Z = 0.0000

Element A37, Line, Intersection calibrator fixture plane and receiver mount plane
179: LINE / INTERSECTION A16 A27
Align z axis to this line
180: ALIGN / AXIS Z_AXIS LAST_ELEMENT

Element A38, Point, intersection of line between mount holes and calibrator
181: POINT / INTERSECTION YZ_PLANE A33
Make this the origin of y-z plane (x is perpendicular to calibrator)
182: ALIGN / ORIGIN YZ LAST_ELEMENT
Print out the following elements
183: PRINT_ELEMENT / ON ALL_ELEMENTS
Turn off projection
184: PROJECTION / OFF
Position of receiver mount hole
185: RECALL / M1
Position of transmitter mount hole
186: RECALL /M3

Element A39, Point, intersection of receiver beam with calibrator
187: POINT / INTERSECTION YZ_PLANE M2

Element A40, Point, Intersection of transmitter beam with calibrator
188: POINT /INTERSECTION YZ_PLANE M4
Plane of receiver mount top
189: RECALL / A5

```
Plane of transmitter mount top
190: RECALL / A16
Stop saving data in file
191: STORE_ASCII / OFF
192: END
```

A.2 DISK STORAGE FORMAT

A.2.1 Original Content

```
N0025   A1    PLANE    MEASURE
        X        2.0178
        Y       44.4255
        Z        4.0774
        PHI    -18:34:52
        THTA    89:59:46
N0027   A2    LINE     MEASURE
        X       -0.9772
        Y        0.1023
        Z       -0.0000
        PHI     98:17:17
        THTA    90:00:00
N0030   A3    LINE     MEASURE
        X       -0.0490
        Y        2.1879
        Z        0.0000
        PHI      0:06:34
        THTA    90:00:00
N0031   A4    POINT    INTERSECTION
        X       -0.9523
        Y        2.1862
        Z       -0.0000
N0034   A5    PLANE    MEASURE
        X        0.9912
        Y       -1.6703
        Z        0.0003
        PHI    177:33:42
        THTA     0:07:13
        RNG      0.0048
```

```
###############################################################################
###############################################################################
#                    HANDAR INC.                                              #
#                    1288 REAMWOOD AVE.                                       #
#                    SUNNYVALE, CA 94089                                      #
#                    TEL. (408) 734-9646                                      #
#                    FAX. (408) 745-7921                                      #
===============================================================================
| DATE         | 7/21/95      | Program name | H:FORK_1D      |
-------------------------------------------------------------------------------
| TIME         | 14:36:12     | OPERATOR     | M. YOUNG       |
===============================================================================
| SERIAL NUMBER | 234         | PART NUMBER  | FAA-10268\1    |
-------------------------------------------------------------------------------
```

REV. LETTER		PART NAME	FORK ASSY
COMMENTS		COMMENTS	
COMMENTS		COMMENTS	

```
   +++ FLATNESS OF PLANE ON LONG FORK. +++
N0040   A5    TOLERANCE
        FLAT     0.0048      0.0100                                      I**--I
N0041   A6    CIRCLE   MEASURE   OUT
        X       -0.6819
        Y       -1.8890
        Z       -0.0000
        DIA      0.2499
        RNG      0.0000
N0042   A7    CIRCLE   MEASURE   OUT
        X        0.7169
        Y        1.4136
        Z       -0.0000
        DIA      0.2499
        RNG      0.0000
   +++ DISTANCE BETWEEN PINS ON LONG FORK. +++
N0044   A8    DISTANCE
        DST      3.5865
N0045   A8    TOLERANCE
        DST      3.5865      3.5850    0.0060   -0.0060    0.0015  I----.*---I
N0048   A9    POINT    BISECTOR
        X       -0.7004
        Y       -1.6509
        Z       -0.0000
N0051   A10   POINT    KEYIN
        X       -0.0000
        Y        0.0000
        Z        0.0000
N0053   A11   LINE     KEYIN
        X       -0.0000
        Y        0.0000
        Z        0.0000
        PHI    -90:00:00
        THTA    90:00:00
   +++ MANUALLY MEASURE 3 POINTS +++
N0061   A12   PLANE    MEASURE
        X       -5.8107
        Y      -39.7830
        Z       12.1094
        PHI     82:32:19
        THTA    38:03:53
   +++ CAUTION CNC MODE WILL START +++
N0067   A13   LINE     MEASURE
        X       -1.0429
        Y        0.1154
        Z        0.0000
```

```
              PHI      69:43:03
              THTA     90:00:00
      N0070   A14   LINE    MEASURE
              X         0.1786
              Y         1.4668
              Z         0.0000
              PHI       0:07:07
              THTA     90:00:00
      N0071   A15   POINT   INTERSECTION
              X        -1.0182
              Y         1.4644
              Z         0.0000
      N0075   A16   PLANE   MEASURE
              X         0.9895
              Y        -2.0772
              Z         0.0012
              PHI    -168:47:02
              THTA      0:04:23
              RNG       0.0055
         +++ FLATNESS OF PLANE ON SHORT FORK. +++
      N0077   A16   TOLERANCE
              FLAT      0.0055      0.0100                                            I***-I
      N0078   A17   CIRCLE  MEASURE  OUT
              X         1.7110
              Y        -3.5620
              Z         0.0000
              DIA       0.2499
              RNG       0.0001
      N0079   A18   CIRCLE  MEASURE  OUT
              X         0.3109
              Y        -0.2602
              Z         0.0000
              DIA       0.2499
              RNG       0.0000
         +++ DISTANCE BETWEEN PINS ON SHORT FORK. +++
      N0081   A19   DISTANCE
              DST       3.5864
      N0082   A19   TOLERANCE
              DST       3.5864      3.5850    0.0060    -0.0060    0.0014  I----.*---I
      N0085   A20   POINT   BISECTOR
              X        -0.7003
              Y         1.6508
              Z         0.0000
      N0088   A21   POINT   KEYIN
              X         0.0000
              Y         0.0000
              Z         0.0000
      N0090   A22   LINE    KEYIN
              X         0.0000
              Y         0.0000
              Z         0.0000
              PHI      90:00:00
```

```
        THTA     90:00:00
N0092   A23   ANGLE
        ANG      42:04:33
   +++ COMPOUND ANGLE BETWEEN PLATES. +++
N0094   A23   TOLERANCE
        ANG      42:04:33    42:00:00    0:15:00    -0:15:00        0:04:33    I----.**--I
N0095   A24   LINE    INTERSECTION
        X         -3.0640
        Y         20.1600
        Z          0.0024
        PHI    -169:03:30
        THTA     90:04:23
N0097   M4    LINE    RECALL
        X         -0.0063
        Y        -20.3751
        Z          0.0000
        PHI      90:00:00
        THTA     90:00:00
N0099   A25   LINE
        X          6.6275
        Y        -19.2671
        Z          0.8183
        PHI     -71:00:00
        THTA     79:03:30
N0100   A25   LINE    RECALL
        X          6.6275
        Y        -19.2671
        Z          0.0000
        PHI     -71:00:00
        THTA     90:00:00
N0101   A11   LINE    RECALL
        X          8.0846
        Y         23.3750
        Z          0.0000
        PHI    -109:04:43
        THTA     90:00:00
   +++ PROJECTED ANGLE BETWEEN THE PLATES. +++
N0103   A26   ANGLE
        ANG      38:04:43
N0104   A26   TOLERANCE
        ANG      38:04:43    38:00:00    0:30:00    -0:30:00        0:04:43    I----.*---I
N0109   A27   PLANE   MEASURE
        X          1.6671
        Y          0.3698
        Z          1.0811
        PHI     -89:24:40
        THTA     89:42:54
        RNG        0.0008
N0110   A28   POINT   INTERSECTION
        X         -0.0000
        Y          0.3472
        Z         -0.0000
```

```
N0113   A29    POINT    INTERSECTION
        X           0.0751
        Y           0.0000
        Z          -3.1668
N0114   A30    POINT    INTERSECTION
        X          -0.0815
        Y          -0.0000
        Z          -3.1655
  +++ THIS IS THE DISTANCE BETWEEN THE INTERSECTION +++
  +++ POINTS. +++
N0117   A31    DISTANCE
        DST         0.1567
N0121   A32    LINE    KEYIN
        X          -0.0000
        Y           0.0000
        Z           0.0000
        PHI         0:00:00
        THTA        0:00:00
N0123   A33    LINE    CONNECT
        X           7.3560
        Y           1.8367
        Z           0.4046
        PHI        88:02:35
        THTA       91:06:40
N0126   A34    LINE    INTERSECTION
        X           0.0724
        Y          -1.6276
        Z           0.0000
        PHI         0:00:00
        THTA        2:32:46
  +++ PROJECTED ANGLE BETWEEN THE LONG FORK AND THE +++
  +++ CENTERLINE. +++
N0130   M2     LINE    RECALL
        X           0.0000
        Y          21.3375
        Z           0.0000
        PHI      -171:12:34
        THTA       90:00:00
N0131   A35    LINE
        X           0.0000
        Y          21.3375
        Z           0.0000
        PHI         8:47:26
        THTA       90:00:00
N0132   A35    TOLERANCE
        PHI         8:47:26      8:42:00   1:00:00  -1:00:00      0:05:26   I----.*---I
  +++ PROJECTED ANGLE BETWEEN THE SHORT FORK AND THE +++
  +++ CENTERLINE. +++
N0134   M4     LINE    RECALL
        X          -0.0000
        Y         -21.3375
        Z           0.0000
```

```
              PHI      -10:35:07
              THTA      90:00:00
    N0135    M4     TOLERANCE
              PHI      -10:35:07   -10:30:00   1:00:00   -1:00:00   -0:05:07   I---*.----I
       +++  _____page 2 of 2_____  +++
       +++  _____END OF TEST_____  +++
```

The following parameters were extracted from the original portion of the stored data:

Element	Description
A5	Flatness of receiver mount
A8	Distance between receiver alignment pins
A16	Flatness of transmitter mount
A19	Distance between transmitter alignment pins
A23	Compound scattering angle
A26	Projection of scattering angle in plane of fork
A31	Distance between intercepts of receiver and transmitter breams at calibrator
A35	Receiver out-of-fork-plane angle
M4/A22	Transmitter out-of-fork-plane angle

A.2.2 Added July 1995

```
N0139    A36    PLANE
         X         -5.6281
         Y         -1.8903
         Z          0.6399
         PHI       0:49:33
         THTA     92:32:45
         RNG       0.0008
N0142    A37    LINE    INTERSECTION
         X          0.1300
         Y          1.7871
         Z         -0.3768
         PHI     -90:08:39
         THTA     90:00:00
N0144    A38    POINT   INTERSECTION
         X          0.0000
         Y          5.6267
         Z          0.6523
N0148    M1     POINT   RECALL
         X         23.0732
         Y          1.0252
         Z          0.3348
N0149    M3     POINT   RECALL
         X        -19.5552
         Y         -0.8689
         Z         -0.2838
N0150    A39    POINT   INTERSECTION
         X          0.0000
         Y         -7.2325
         Z          3.5957
N0151    A40    POINT   INTERSECTION
         X          0.0000
```

```
              Y          -7.3782
              Z           3.5887
    N0152   A5    PLANE    RECALL
              X          23.0121
              Y           1.0034
              Z           0.3260
              PHI         0:11:48
              THTA      109:40:02
              RNG         0.0048
    N0153   A16   PLANE    RECALL
              X         -19.4256
              Y          -0.9108
              Z          -0.2799
              PHI        -0:00:00
              THTA       71:36:13
              RNG         0.0055
```

The following parameters were extracted from the added portion of the stored data:

Element	Description
M1	Position of receiver mount hole: X, Y, Z
M3	Position of transmitter mount hole: X, Y, Z
A39	Intercept of receiver beam with calibrator: Y, Z
A40	Intercept of transmitter beam with calibrator: Y, Z
A23	Compound scattering angle
A26	Projection of scattering angle in plane of fork
A5	Receiver mount top plane: PHI, THTA
A16	Transmitter mount top plane: THTA
M4/A22	Transmitter out-of-fork-plane angle

A.3 HARD COPY FORMAT

A.3.1 Original

```
###############################################################################
###############################################################################
#                   HANDAR INC.                                                #
#                   1288 REAMWOOD AVE.                                         #
#                   SUNNYVALE, CA 94089                                        #
#                   TEL. (408) 734-9646                                        #
#                   FAX. (408) 745-7921                                        #
===============================================================================
| DATE          | 7/21/95         | Program name | H:FORK_1D            |
-------------------------------------------------------------------------------
| TIME          | 14:36:12        | OPERATOR     | M. YOUNG             |
===============================================================================
| SERIAL NUMBER | 234             | PART NUMBER  | FAA-10268\1          |
-------------------------------------------------------------------------------
| REV. LETTER   |                 | PART NAME    | FORK ASSY            |
-------------------------------------------------------------------------------
| COMMENTS      |                 | COMMENTS     |                      |
-------------------------------------------------------------------------------
| COMMENTS      |                 | COMMENTS     |                      |
```

no	sym	actual	nominal	u_tol/tp	l_tol/edp	dev	out of tol

+++ FLATNESS OF PLANE ON LONG FORK. +++

| N0040 | A5 | TOLERANCE | | | | | |
| | FLAT | 0.0048 | 0.0100 | | | | I**--I |

+++ DISTANCE BETWEEN PINS ON LONG FORK. +++

| N0045 | A8 | TOLERANCE | | | | | |
| | DST | 3.5865 | 3.5850 | 0.0060 | -0.0060 | 0.0015 | I----.*---I |

+++ FLATNESS OF PLANE ON SHORT FORK. +++

| N0077 | A16 | TOLERANCE | | | | | |
| | FLAT | 0.0055 | 0.0100 | | | | I***-I |

+++ DISTANCE BETWEEN PINS ON SHORT FORK. +++

| N0082 | A19 | TOLERANCE | | | | | |
| | DST | 3.5864 | 3.5850 | 0.0060 | -0.0060 | 0.0014 | I----.*---I |

+++ COMPOUND ANGLE BETWEEN PLATES. +++

| N0094 | A23 | TOLERANCE | | | | | |
| | ANG | 42:04:33 | 42:00:00 | 0:15:00 | -0:15:00 | 0:04:33 | I----.**--I |

+++ PROJECTED ANGLE BETWEEN THE PLATES. +++

| N0104 | A26 | TOLERANCE | | | | | |
| | ANG | 38:04:43 | 38:00:00 | 0:30:00 | -0:30:00 | 0:04:43 | I----.*---I |

+++ THIS IS THE DISTANCE BETWEEN THE INTERSECTION +++
+++ POINTS. +++

| N0117 | A31 | DISTANCE | | | | | |
| | DST | 0.1567 | | | | | |

no	sym	actual	nominal	u_tol/tp	l_tol/edp	dev	out of tol

+++ PROJECTED ANGLE BETWEEN THE LONG FORK AND THE +++
+++ CENTERLINE. +++

| N0132 | A35 | TOLERANCE | | | | | |
| | PHI | 8:47:26 | 8:42:00 | 1:00:00 | -1:00:00 | 0:05:26 | I----.*---I |

+++ PROJECTED ANGLE BETWEEN THE SHORT FORK AND THE +++
+++ CENTERLINE. +++

| N0135 | M4 | TOLERANCE | | | | | |

```
       PHI     -10:35:07   -10:30:00    1:00:00   -1:00:00    -0:05:07   I---*.----I
-------------------------------------------------------------------------------
  +++ _____page 2 of 2_____  +++
  +++ _____END OF TEST_____  +++
```

A.3.2 Added July 1995

```
-------------------------------------------------------------------------------
N0148   M1    POINT    RECALL
        X       23.0732
        Y        1.0252
        Z        0.3348
N0149   M3    POINT    RECALL
        X      -19.5552
        Y       -0.8689
        Z       -0.2838
N0150   A39   POINT    INTERSECTION
        X        0.0000
        Y       -7.2325
        Z        3.5957
N0151   A40   POINT    INTERSECTION
        X        0.0000
        Y       -7.3782
        Z        3.5887
N0152   A5    PLANE    RECALL
        X       23.0121
        Y        1.0034
        Z        0.3260
        PHI      0:11:48
        THTA   109:40:02
        RNG      0.0048
N0153   A16   PLANE    RECALL
        X      -19.4256
        Y       -0.9108
        Z       -0.2799
        PHI     -0:00:00
        THTA    71:36:13
        RNG      0.0055
```

APPENDIX B - SIMCAL8 DETAILS

Section 3.4 presented the assumptions of the simulation and the input and output parameters. This appendix describes the details of the calculation.

B.1 HEAD GEOMETRY

The first step in the calculation is to take into account the displacements of the beam origins and calibrator intercepts by the head geometry parameters (see Figure 15). This process is done in three steps:

1. The beam origins are displaced along the beam vectors by the parameters d_t and d_r. The calibrator intercepts are unchanged.

2. The beam origins and calibrator intercepts are displaced along the normal vector to the mounting plate by the amounts o_t and o_r.

3. The calibrator intercepts then are moved along the beam vectors back to x=0, which is the correct intercept.

These displacements were validated by the reasonableness of the changes in position and the constant beam vector pointing before and after the displacements. In this validation process, several errors were found.

B.2 SCATTERING CALCULATION

The scattering calculation is carried out in planes of fixed x coordinate. Only one plane (x=0) is calculated for the plate scattering. Many planes are calculated for the volume scattering. Volume scattering is added for values of x farther and farther away from zero until no overlap is found between the transmitter and receiver beams.

Within an x plane, the smaller beam (selected as the receiver for x>0 and the transmitter for x≤0) is searched completely for points that also are included in the other beam. The algorithm for searching the plane is designed to make sure that no points are overlooked. The intersection of a conical beam with a plane is an ellipse with the major axis oriented near the y axis. The scan pattern is shown in Figure 71. The middle of the ellipse is scanned horizontally (along z axis) until the point is reached on the top and bottom where the farthest z points reached on both sides of the ellipse are

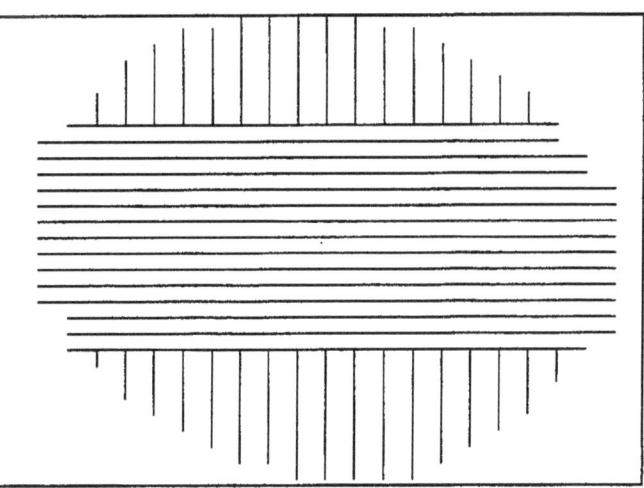

Figure 71. Pattern for Scanning Ellipse

less than their extreme values. The top and bottom sections are then scanned vertically (along y axis). Figure 71 was actually developed for earlier simulations; the ellipse for SIMCAL8 actually is rotated by 27 degrees with respect to the y-z coordinate system. Checking the operation of the code showed that the algorithm, nevertheless, works properly with a rotated ellipse.

REFERENCES

[1] Burnham, D.C., Spitzer, E.A., Carty, T.C., and Lucas, D.B., "United States Experience Using Forward Scattermeters For Runway Visual Range," Report No. DOT/FAA/AND-97/1, March 1997.

[2] Griggs, D.J., Jones, D.W., Ouldridge, M., and Sparks, W.R., "The First WMO Intercomparison of Visibility Measurements, United Kingdom 1988/1989," Final Report WMO/TD-No. 401, 1990.

[3] Burnham, D.C. and Pawlak, R.J., "Accuracy Requirements for Runway Visual Range Systems," Poster Paper P13.11, Eighth Conference on Aviation, Range, and Aerospace Meteorology, January 10-15, 1999, Dallas TX, American Meteorological Society.

[4] FAA Runway Visual Range Specification, FAA-E-2772.

[5] West, M.D., D. C. Burnham, and C. S. Miles, "Calibration Error Simulation for Forward Scatter Visibility Sensors," Sixth Conference on Aviation Weather Systems, January 15-20, 1995, Dallas, TX, American Meteorological Society, pp. 341-346.

[6] Burnham, D., Pawlak, R, Carty, T., Lucas, D, and Collett, S, "Joint US-UK Runway Visual Range Tests: 1994-1999, to be published.

[7] McCartney, E.J. *Optics of the Atmosphere: Scattering by Molecules and Particles*, Wiley, New York, NY, 1976.

www.ingramcontent.com/pod-product-compliance
Lightning Source LLC
Chambersburg PA
CBHW081830170526
45167CB00007B/2771